Bayesian A/B
Decision Models

Bayesian A/B
Decision Models

J Christopher Westland

University of Illinois Chicago, USA

World Scientific

NEW JERSEY · LONDON · SINGAPORE · BEIJING · SHANGHAI · HONG KONG · TAIPEI · CHENNAI · TOKYO

Published by

World Scientific Publishing Co. Pte. Ltd.
5 Toh Tuck Link, Singapore 596224
USA office: 27 Warren Street, Suite 401-402, Hackensack, NJ 07601
UK office: 57 Shelton Street, Covent Garden, London WC2H 9HE

Library of Congress Cataloging-in-Publication Data
Names: Westland, J. Christopher, author.
Title: Bayesian A/B decision models / J. Christopher Westland, University of Illinois Chicago, USA.
Description: Hackensack, NJ : World Scientific, [2025] |
    Includes bibliographical references and index.
Identifiers: LCCN 2024041763 | ISBN 9789819802036 (hardcover) |
    ISBN 9789819802043 (ebook for institutions) | ISBN 9789819802050 (ebook for individuals)
Subjects: LCSH: Bayesian statistical decision theory.
Classification: LCC QA279.5 .W48 2025 | DDC 519.5/42--dc23/eng/20250208
LC record available at https://lccn.loc.gov/2024041763

British Library Cataloguing-in-Publication Data
A catalogue record for this book is available from the British Library.

For any available supplementary material, please visit
https://www.worldscientific.com/worldscibooks/10.1142/14077#t=suppl

Desk Editors: Murali Appadurai/Analyn Alcala

Typeset by Stallion Press
Email: enquiries@stallionpress.com

Preface

Bayesian A/B decision models have evolved significantly in statistical analysis and decision-making, especially over the last decade. This preface introduces my book, *Bayesian A/B Decision Models*, which encapsulates my experience utilizing these models in highly specialized areas such as auditing and as an expert witness in computing damages for corporate and industrial business legal cases.

My journey began with a recognition that, while traditional statistical methods provided valuable insights, they often fell short in addressing complex, real-world problems where nuanced decision-making is crucial. The Bayesian A/B decision approach, although initially daunting due to its complexity and the steep learning curve, offered a robust alternative. The startup costs — both in time and effort — to master this advanced statistical tool are nontrivial, yet the benefits it brought to my professional capabilities were well worth the effort.

As a consultant and legal expert, the flexibility of Bayesian A/B models proved invaluable. These models allowed me to respond adeptly to diverse and dynamic requests for detailed analytical reports. I found myself capable of producing specific financial analyses and outcomes that were not only highly tailored but also often unattainable by the opposing side using more traditional methods. This edge was instrumental in numerous legal and corporate deliberations, where precision and adaptability of the statistical analysis could make or break a case.

Over the years, my engagement with Bayesian A/B models has evolved, revealing applications that spanned various sectors and challenges. This book is not merely a reflection of my own experiences but also serves as a guide to applying Bayesian decision models across nearly every imaginable scenario. The examples included range from healthcare, marketing, and finance to more niche sectors where traditional A/B testing or frequentist approaches might falter due to their inherent limitations.

One of the unique aspects of Bayesian A/B testing is its requirement for a powerful, flexible statistical language that can handle the complexity and nuances of Bayesian analytics. Throughout my practice, I have found the *R statistical language* to be a singularly effective and efficient tool for the implementation of my ideas and models. It not only supports the necessary statistical rigor but also provides a versatile platform for developing, testing, and deploying Bayesian models.

To this end, I have included a comprehensive technical appendix in this book, which contains a complete set of R code for all the examples discussed. This resource aims to not just illustrate the theoretical application of Bayesian A/B tests but to also provide practical tools that can be adapted and implemented by readers in their own fields. For those who prefer a digital format or wish to interact with the code directly, all resources are available on my GitHub site at `https://github.com/westland/BayesAB`. Whether you are a seasoned statistician, a budding analyst, or a professional curious about the potential of Bayesian A/B testing, this book offers a comprehensive overview and detailed applications that demonstrate the unparalleled benefits of this approach.

I sincerely hope that this book not only enlightens you about the sophisticated world of Bayesian A/B testing but also encourages you to explore its potential in your own areas of expertise. The journey from a novice to a proficient user of Bayesian models is challenging yet rewarding, and it is a path I recommend enthusiastically to all who are committed to excellence in decision-making and statistical analysis.

J. Christopher Westland, PhD, CPA
Chicago, Illinois, USA
July, 2024

About the Author

J. Christopher Westland is Professor in the Department of Information & Decision Sciences at the University of Illinois Chicago. He has a BA in Statistics and an MBA in Accounting from Indiana University and received his PhD in Computers and Information Systems from the University of Michigan. He has professional experience in the US as a certified public accountant and as a consultant in technology law in the US, Europe, Latin America, and Asia. He has authored numerous academic papers and scholarly books including *Audit Analytics: Data Science for the Accounting Profession*, *Global Innovation Management*, *Red Wired: China's Internet Revolution*, *Global Electronic Commerce*, *Structural Equation Modeling*, *Financial Dynamics*, and *Valuing Technology*. He is the Editor-in-Chief of *Electronic Commerce Research* journal and serves on the editorial boards of numerous technology journals. He has served on the faculties at the University of Michigan, University of Southern California, Hong Kong University of Science and Technology, Tsinghua University, University of Science and Technology of China, Harbin Institute of Technology, and other academic institutions. He is fluent in Mandarin and, in 2012, received High-Level Foreign Expert status in China under the 1000-Talents Plan.

Contents

Chapter 1

What are A/B Decision Models?

1.1. Claude Hopkins' Legacy

A/B decision models are alternately known as bucket testing, split-run testing, and split testing. They consist of a randomized experiment involving two variants of a particular method, A and B, thus the name A/B testing. These variants can be a "treatment" and a "control", as in an agricultural or laboratory setting. More often, they are simply two different treatments, e.g., two alternative advertising campaigns for a product.

Rudimentary forms of A/B tests have been performed since the 18th century [1]. The modern popularity of the method is mainly attributable to advertising pioneer Claude Hopkins, who used promotional coupons, which could be tracked and counted, to test the effectiveness of his advertising campaigns. Eventually, Hopkins's scientific advertising [2] methods were augmented with statistical methods developed by William Sealy Gosset [3] to create our traditional frequentist method of A/B testing.

The move to scientific advertising began with a piqued retailer. Around the turn of the 20th century, the department store tycoon John Wanamaker complained that "half of my advertising budget is wasted; the problem is that I don't know which half". Hopkins' innovations allowed retailers like Wanamaker to track the effectiveness of their advertising expenditures more precisely.

Hopkins's work notably promoted Bissell Carpet Sweepers, Swift sausages, and Dr. Shoop's patent medicines. Hopkins was not above controversy in the pursuit of effective advertising. The government

1

shut down Dr. Shoop's for selling useless medicines, but not before earning significant sales from Hopkins' advertising. Hopkins's sensationalist campaigns hawked Schlitz beer by highlighting the company's use of "live steam" for cleaning bottles — a common practice among all brewers, not just Schlitz. He claimed that Cleopatra bathed with Palmolive soap, invented only in 1898. He said that Puffed Wheat was shot from guns to puff the grains to eight times their normal size, a claim that, this time in contrast to his many other claims, was true.

Hopkins insisted on thorough research into the client's products before crafting a campaign. That involved repeated questioning of why a customer would purchase the product. He supported representative sampling of consumer feedback — what would be called "stated preferences" today — followed by an analysis of promotional coupon refunds to ascertain the "revealed preferences" of the consumers. Several successes of his A/B testing approach are notable. Bissell's direct mail campaign promoted carpet sweepers as ideal Christmas gifts, resulting in a staggering response of one thousand orders from just five thousand letters [4]. His greatest successes were with Pepsodent toothpaste. Hopkins initially did not want Pepsodent as a client, feeling that it would not be successful due to lack of demand and numerous competitors, but ended up taking them as a client as a favor to a friend. In the early 20th century, infections from poor dental hygiene had been the major reason for rejection from US military service in World War I. Within five years, his efforts helped make Pepsodent one of the most recognized brands globally, celebrated by figures like Shirley Temple and Clark Gable for their "Pepsodent smiles" and brushing had become a regular habit with Americans [5].

Claude Hopkins' successful use of A/B testing transformed advertising. His followers further capitalized on emerging statistical methods from brewing, agriculture, psychology and the social sciences — the hypothesis test. Hypothesis testing would later become a cornerstone in A/B decision models, illustrating a fascinating interplay between statistics and consumer behavior.

Sensationalist though he was, Hopkins discipline and innovation had a lasting effect on not only marketing, but also many other fields. This was the legacy of A/B decision models. It is this legacy that I explore in the current book.

1.2.   What is A/B Testing?

A/B testing is a statistical decision-making process that employs a simple randomized controlled experiment, often involving the dichotomization of user choices. A/B testing is a shorthand for a simple randomized controlled experiment in which a number of samples of a single vector-variable are compared. A/B tests are widely considered the simplest form of controlled experiment, especially when they only involve two variants. They use the same methodologies as simple hypothesis tests, from which they borrow their mathematical form. In broader scenarios, such as bandit A/B models, decision options are referred to as "arms," akin to the arms of a series of slot machines, where decisions can extend beyond just two choices.

A/B tests are a subset of statistical hypothesis testing methods. These are mathematical procedures that determine whether the observations in a dataset are consistent with one or the other of two alternative hypotheses. For simplicity, we will call these alternative hypotheses A and B. The process begins with distilling the model and dataset into a so-called "test statistic," which is then compared to some predetermined "cutoff" value. Ideally, this test statistic will be directly interpretable and measured in the same units as the decision objective. Usually, though, this is not the case due to limitations of the statistical methods. Over the years, approximately 100 specialized statistical tests have been developed, greatly enhancing the tools available to researchers and marketers.

Hypothesis testing has historical roots dating back to the 1700s when John Arbuthnot and later Pierre–Simon Laplace studied human sex ratios at birth. In the 20th century, Karl Pearson (who introduced the much-maligned p-value and chi-squared test) and Ronald Fisher defined the "null hypothesis," and the "significance test". The formal framework of hypothesis testing was developed by Jerzy Neyman and Egon Pearson, culminating in the Neyman–Pearson Lemma [6,7].

Despite its simplicity, A/B testing has developed a complex and specialized vocabulary specific to its use in various fields, such as marketing and clinical trials. This complexity often arises when non-statisticians attempt to navigate the inherently ambiguous and cumbersome frequentist approach to A/B testing.

1.3.   Frequentists and Bayesians Approach A/B Decision Models

Traditionally A/B testing has borrowed the Neyman–Pearson machinery of hypothesis testing developed in the 1930s to generate a test statistic, the p-value, that measures whether alternative A or B is to be preferred. This is the simple one-step process depicted in Figure 1.1.

The p-value is the probability of obtaining test results at least as extreme as that observed, under the assumption that decision A is preferred to B. Software to quickly transform any dataset into a p-value is readily available but is widely misunderstood and misused. Furthermore, like other frequentist approaches, the p-value is only an objective measure of our preference for A over B for an infinite sample. Even though reporting p-values of statistical tests is common practice in academic publications of many quantitative fields, misinterpretation and misuse of p-values is widespread. This has become a significant topic in mathematics and meta-science [8–10]. In 2016, the American Statistical Association issued a formal statement that "p-values do not measure the probability that the studied hypothesis is true, nor the probability that the data were produced by random chance alone" and that "a p-value, or statistical significance, does not measure the size of an effect or the importance of a result" or "evidence regarding a model or hypothesis" [11]. More importantly, virtually all of the questions asked in A/B testing are measured in something besides a probability score restricted to the range $p \in [0, 1]$. For example, advertising tests are typically measured in increased customer spending, profitability, click-through rates on websites, and so forth. It isn't easy to see how even a correctly calculated p-value can be transformed into one of these figures of merit [8].

There are also conceptual problems in the frequentist machinery. Limitations in their definition of probability add complexity and require the researcher to conduct power tests, with some arbitrary

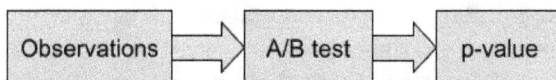

Figure 1.1.   Components of a frequentist A/B Test.

assumptions about the minimum acceptable power, to ascertain the necessary sample size. The statistics and methods used can be controversial, and their practice is somewhat of a black art. Where data is costly, sample size limitations make frequentist decisions brittle and unreliable, changing with every acquisition of new data.

The Bayesian A/B decision process, depicted in Figure 1.2, has its detractors. It has been criticized as complex and computationally expensive. You can see from Figure 1.2 that Bayesian approaches are more complex; this is true. But it is because they incorporate more sources of information. Computation expense was a concern 20 years ago, but new algorithms and rapid advances in hardware now make Bayesian A/B methods accessible on even the cheapest microcomputer. Let's take a closer look at the components in Figure 1.2.

Like their traditional frequentist counterparts, Bayesian A/B tests start by planning and conducting randomized experiments or quasi-experiments to collect data for inference. However, they also reflect that in any situation, we know a lot about factors influencing the decision even before the first observation in the dataset is collected. Consider an A/B test in which we are trying out a modification, option B to the landing page on our website, versus option A of keeping the current landing page unchanged. Let our figure of merit be the number of visits per month. If we have averaged 1,500 visits per month over the past year, with a maximum of 2,000 and a minimum of 1,000, then we would assign a low probability to next month receiving 10,000 visits under either option A or B. Thus we could choose a prior distribution $x \sim Normal(1500, 500^2)$

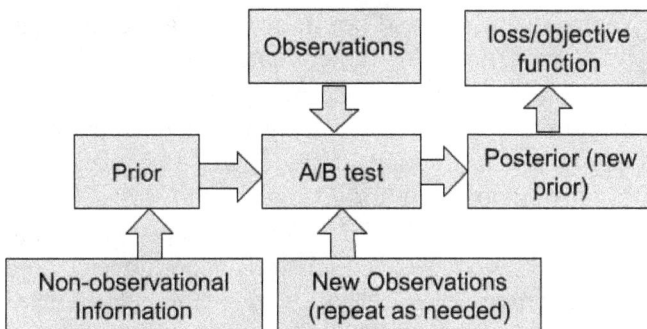

Figure 1.2. Components of a bayesian A/B test.

visits to incorporate knowledge from our historical visits. If we knew that the coming month would feature a widely advertised promotion which we think can attract another 200 visits per month, we might adjust our prior to be $x \sim Normal(1700, 500^2)$. Sure, this is subjective, but we have a chance to learn from the observations in our dataset whether our prior beliefs are accurate or biased. The frequentist counterpart to the Bayesian prior is an assumption that we have no information at all about the world outside the dataset — that is $x \sim Normal(0, \infty)$. This is not only untrue, but it purposely wastes useful information that could help us avoid unnecessary costs of data collection, and potentially could lower our sample sizes.

Just as in frequentist A/B testing, the observations in the dataset are fed to the A/B methods statistical machinery. But rather than outputting a p-value, they are used to generate a likelihood function. This function does exactly what its name implies — it computes the likelihood of the dataset it received over all of the possible values of the posterior distributions parameters, e.g., the mean and standard deviation of a Normal distribution. The likelihood and prior distribution are combined then to create the posterior. In the examples in this book, the priors and posteriors are what are called "conjugate" — the prior and posterior are the same distribution and the likelihood does not change that. The likelihood function encapsulates *what we have learned from the data*; the prior distribution encapsulates *what we know about world in advance of data collection.*

The posterior distribution adjusts our prior beliefs, corrected for what we have discovered about the world through the observations in our dataset. Bayesians do not need infinite datasets; they are willing to admit their limitations and see data as a way to learn more.

Once we have a posterior distribution, we can convert it into a meaningful form for our decision by applying a loss or objective function. For example, if we are measuring revenue from website visits, we could apply a loss function of the form "each visit generates \$4 of revenue". If our posterior for B is $x \sim Normal(2000, 500^2)$ then $revenue \sim Normal(\$8000, \$2000^2)$ and our expected revenue would be \$8,000, a \$2,000 increase per month. But perhaps we are less concerned about average revenue than we are about keeping the minimum number of sales above 1000 unit per month. Then we might be interested in the probability that we will have fewer than 1,000 sales. For this we compute the probability in the lower tail below 1,000 for $Normal(2000, 500^2)$; this probability is equal to 2.28%. Neither of these results is easily obtained from frequentist A/B tests; both are the sort of results that are easy to communicate and are directly relevant to an organization's operations.

The sophistication of Bayesian methods comes with the cost of increased complexity. However, recent advancements in computing and software have made Bayesian approaches to A/B testing almost as straightforward as traditional methods, providing more robust and informative outcomes without significantly adding complexity.

1.4.   Notable A/B Testing Successes

Since Claude Hopkins' Scientific Advertising, A/B testing has become a mainstay of marketing and an essential tool in optimizing online advertising campaigns. Today, tech companies like Microsoft and Google each conduct over 10,000 A/B tests annually. Here are a few examples of how A/B tests have yielded benefits in the recent past.

Barack Obama's presidential campaign: In 2007, Barack Obama's presidential campaign used A/B testing as a way to garner online attraction and understand what voters wanted to see from the presidential candidate [24]. For example, Obama's team tested four distinct buttons on their website that led users to sign up for newsletters. Additionally, the team tested six different images to draw in users. Through A/B testing, staffers were able to determine how to draw in voters and garner additional interest effectively.

Kiva microfunds: Kiva is a non-profit headquartered in San Francisco, California whose mission is to expand financial access to help underserved communities thrive. Kiva distributes funds that it receives to microfinance institutions, social impact businesses, schools or non-profit organizations and does not generally directly provide funds to specific individuals. Kiva conducted an A/B Test as they wanted to increase the number of donations from first-time visitors to their landing page. Donations increased by 11.5% after adding an information box at the bottom of the landing page. This altered the landing page in a way designed to answer any questions that a visitor might have.

Designhill: Mentioning the title of the blog in the subject line of the email would get the majority of click-through rather than requesting recipients to review the post with the blog's title. Through this change the company was able to score 5.84% higher CTR and 2.57% higher open rate by including just the title of the blog in the subject line.

Swiss Gear: Swiss Gear experimented with their product page to increase their conversions. The original version of the product page colors were dominated by red and black, which didn't result in one specific part standing out from the rest of the page. They tested alternatives, highlighting the most important elements in red to make them stand out from the rest of the page. The "special price" and "add to cart" sections immediately caught customers' attention compared to the original version. This small change increased conversions by 132%.

Sony VAIO: Sony experimented with the banner of their VAIO laptops' ad to see what works best for their campaign and which version had the highest shopping cart adds. The results of their A/B tests increased the click through rate by 6% compared with the control version, and the shopping cart advertising increased by 21.3%.

Microsoft: A/B testing on the Bing search engine explored alternatives for displaying advertising headlines. Within hours, the alternative format produced a revenue increase of 12% with no impact on user-experience metrics.

Codecademy: Codecademy provides free coding classes in 12 different programming languages. They A/B tested the application of the "Rule of 100" — that customers see amounts over $100 as being greater in value than percentages, even when both represent the same dollar amount. Their test resulted in an increase of 28% in annual pro plans and an increase in overall page conversions.

Insightsquared: Insightsquared is a B2B sales forecasting platform. They noticed that only 15% of the users left their phone numbers, and A/B tested whether this field could be dropped. Dropping the field resulted in an increase of 112% in conversions.

Århus Teater: Århus Teater is one of Denmark's biggest and oldest theaters. They A/B tested several versions of their "call to action" button on their website. By changing the call to action from "ticket" to "tickets" the theater increased its click through rate by 49% and its ticket sales by 20%.

Rasmussen College: Rasmussen College, a for-profit private college, wanted to increase leads from Pay-Per-Click traffic on their mobile site. They A/B tested mobile-optimized landing pages and sites, increasing their leads by 256%.

Beckett Simonon: Beckett Simonon is an online brand selling high-quality shoes and accessories with an objective of reducing waste and producing responsibly and ethically. A/B testing showed an increase of 5% when showcasing their values in between their products compared to the control version, where they just showcased their products.

1.5. Bayesian A/B Testing

Traditional A/B testing methodologies primarily rely on frequentist hypothesis tests to generate a point estimate — the probability of rejecting the null hypothesis — of a value that can be challenging to interpret. Typically, statisticians or data scientists must conduct a power analysis to determine the necessary sample size, then collaborate with a Product Manager or Marketing Executive to communicate the results, a process that often complicates interpretability. Moreover, these methods lack the robustness of A/B testing that utilizes informative priors and examines the entire distribution of a parameter rather than just a point estimate.

Bayesian approaches offer significant advantages in A/B testing, particularly in terms of interpretability. Instead of p-values, Bayesian methods yield direct probabilities about whether variant A is better than variant B, and by what margin. Posterior distributions in Bayesian tests are parametrized random variables that can be summarized in numerous ways and are immune to "peeking", making them valid regardless of when a test is halted.

In Bayesian analysis, you define your prior beliefs about a parameter's possible distribution mathematically. As you expose different groups to varying tests, you gather data and merge it with your prior to derive the posterior distribution of the parameter(s) in question.

There are numerous reasons for choosing Bayesian methods for A/B testing and other statistical analyses. Ease of communication and interpretability are crucial, and significantly improved with Bayesian approaches. For instance, it is more straightforward to state, "$P(A > B)$ is 5%", than to say, "Assuming the null hypothesis that A and B are equal, the probability of observing a result as extreme as this in A versus B is 5%". Additionally, Bayesian methods allow statements like, "There is a 95% chance that A's mean is between 25.9 and 42.1", directly derived from the analysis.

Using an informative prior mitigates many common issues in standard A/B testing, such as the risk of repeated testing, where hypothesis tests are recalculated as new data arrives. Ideally, in a frequentist approach, you would determine the sample size via a power test and refrain from examining your data until the required sample is collected. Each analysis risks a false positive, and repeated testing significantly increases the likelihood of such errors. An informative prior ensures that the posterior distribution remains meaningful at any examination point. If the posterior distribution appears incorrect, it may indicate an issue with the initial choice of priors.

Additionally, an informative prior addresses the low base-rate problem, where the probability of a successful observation is inherently low. This makes the posterior distribution more stable from the start.

The current book uses the R package `bayesAB` for its inference and mathematics. The typical `bayesAB` workflow includes the following steps:

- **Data Parameterization:** Decide how to parameterize your data (e.g., Poisson for counts of email submissions, Bernoulli for click-through rates on an ad).
- **Priors Selection:** Use helper functions (such as bayesTest and plotDistributions) to select priors for your data.
- **Perform the Bayes A/B test:** Utilize `print`, `plot.bayesTest`, and `summary.bayesTest` functions to interpret your results. In addition, you can merge multiple `bayesTest` objects for a non-analytical target distribution.
- **Early Test Stopping:** Assess whether to stop your test early based on the Posterior Expected Loss provided in the summary output.

Additionally, you can use the `banditize` and `deployBandit` functions to convert a pre-calculated (or empty) bayesTest into a multi-armed bandit. This can be used for serving recommendations and adapting as new data is collected. Although `bayesAB` is designed for A/B testing data, it can also be used for Bayesian analysis on any data vector, provided it can be parameterized using the available functions.

1.6. Moving Forward

The next two chapters will present topics representing significant factors in the researcher's decision to adopt Bayesian A/B methods.

Chapter 2 explores the limitations and challenges associated with traditional frequentist A/B testing. It illustrates how these methods, while popular, are fraught with pitfalls that can ensnare an unwary researcher. Additionally, even when an A/B test is conducted correctly, the interpretation of the results remains a significant hurdle. The p-values produced by frequentist tests often do not align with the practical decisions an organization needs to make, such as enhancing profitability or increasing click-through rates.

In Chapter 2, I describe how Bayesian A/B testing methods offer an interpretable and flexible framework for decision-making. Bayesian approaches can be precisely tailored to meet the specific needs of decision-makers. This adaptability proves invaluable when explaining results and making decisions in a professional setting, where clear communication and understanding are crucial among stakeholders with diverse agendas. Unlike frequentist methods, which are limited by the ambiguous and often controversial p-value, Bayesian methods provide clarity and adaptability in interpreting results, aligning closely with the goals and language of organizations. Consequently, Bayesian approaches facilitate actionable decisions and establish clear benchmarks for assessing performance.

Chapter 3 addresses the misconceptions surrounding prior distributions in Bayesian analysis. Historically, a significant criticism of priors has been their perceived subjectivity. This chapter challenges this notion by demonstrating that prior distributions can be equally objective and offer greater flexibility compared to the traditional frequentist methods used in A/B testing.

Chapter 2

Twelve Questions You Should Ask in Advance of A/B Testing

A/B testing is a pivotal methodology in the contemporary landscape of business and research, serving as a critical link between theoretical enhancements and evidence-based decision-making. This method, essentially a randomized controlled trial, is widely adopted across various industries to assess alterations in services or products. Its core advantage lies in its straightforward methodology: it contrasts a control group (Group A) with a treatment group (Group B) under specified conditions to determine the impact of changes on user behavior or system performance.

At its heart, A/B testing functions as a scientific experiment to assess the effectiveness of innovations within a controlled setting before broader application. This technique is vital for companies implementing data-driven strategies to improve user experience, interface updates, or content personalization. The methodology involves exposing one segment of users to a new service version while another segment continues using the existing version. The comparative analysis of responses from these groups helps organizations assess whether the new modifications positively affect user engagement or operational outcomes.

A/B testing is adaptable and useful in numerous contexts where direct comparisons yield important insights:

- **User Interaction:** Altering the sequence of information presentation may boost user engagement and increase time on a platform.

- **Content Personalization:** Customizing cover art for streaming content could enhance appeal and elevate consumption levels.
- **System Optimization:** A revised recommendation system may improve user responsiveness and satisfaction.
- **Performance Analysis:** Evaluating the three-point accuracy of two basketball players can expose differences in skill levels.
- **Clinical Research:** In drug trials, A/B testing is essential for determining whether a new treatment is more effective than a placebo.

Traditionally, A/B testing relies on frequentist statistical methods that emphasize hypothesis testing and p-values. However, this approach can lead to misinterpretations and potentially misleading conclusions. The term "statistical significance" is often used indiscriminately, sometimes resulting in decisions based on arbitrary thresholds without considering deeper implications.

The Bayesian methodology offers a sophisticated alternative for A/B testing by incorporating domain-specific knowledge and probabilistic reasoning into decision-making. This approach enhances the understanding of data and outcomes, providing a comprehensive analysis beyond the simplistic binary conclusions of traditional methods. Key benefits of the Bayesian method include the following:

- **Incorporation of Prior Knowledge:** Experts can integrate their insights and expectations into the model, enriching the interpretation of experimental data.
- **Focus on Data Modeling:** Instead of deriving test statistics, Bayesian methods model the data directly, enabling more accurate predictions of outcomes.
- **Probabilistic Interpretation:** Differences between groups are articulated in probabilities, offering an intuitive and flexible understanding of results.
- **Quantification of Null Hypotheses:** Setting priors to reflect null hypotheses creates a framework for comparing experimental data against baseline expectations.

Implementing the Bayesian approach requires setting up a model that predicts outcomes based on group membership and other pertinent factors. Typically, this model assumes a normal distribution for continuous data or a binomial distribution for count data, depending on the metrics under analysis.

A/B decision models can, in theory, provide clear insights into user preferences by directly comparing one option against another. The method relies on actual user behavior rather than stated intentions, making the data especially useful for determining which option performs better. Despite these benefits, A/B decision models have limitations. While they excel at addressing specific design questions, their usefulness is limited to problems with easily measurable outcomes using easily measurable observations. They also require the clear articulation of model parameters, data and objectives, a process that can be costly and time-consuming. Larger organizations may spend significant time and effort to identify and prioritize figures of merit and objectives, and this may involve extensive meetings and discussions, with the potential for wasted time and resources.

Bayesian A/B models inherently support several features that are not addressed by frequentist/traditional A/B tests:

- **Interpretability:** Bayesian methods provide clear and intuitive interpretations of results.
- **Assessment of Value and Money:** Bayesian approaches can directly incorporate cost-benefit analyses.
- **Data and Decisions Over Time:** Bayesian methods naturally handle data and decision-making over time. In contrast, frequentist statistics require that the experimental design be specified in advance. Frequentist methods too often waste information because it is not in the "right" form (i.e., a matrix-style dataset of observations), whereas Bayesian methods can incorporate a wide variety of information formats into the priors:
- **Learning Processes:** Unlike frequentist statistics, Bayesian statistics derive all relevant information from the observed data, without relying on unobserved quantities.
- **Provisional Decisions:** Bayesian methods recognize that all decisions are provisional and can be revised when new data is received, making them less sensitive to rapidly changing situations.
- **Role of Data:** In frequentist approaches, there is ambiguity in whether the method is summarizing or reporting data, or if the decision is data-driven.
- **Exchangeability:** Bayesian methods only require exchangeability, whereas frequentist A/B testing demands the more stringent requirement of independent identically distributed random samples, which is often impractical.

In this chapter, I delineate the main challenges facing the researcher in implementing an A/B test and the main questions that every researcher should ask before beginning their A/B testing, with my commentary on managing these challenges.

2.1. Can You Replicate the Same Decision with New or Updated Data?

Poor replicability in statistical inference is a significant concern in many scientific fields. Replicability refers to the ability of a study's results to be consistently reproduced when the study is repeated under similar conditions. When findings cannot be replicated, it casts doubt on their validity and undermines confidence in the research. Several factors contribute to poor replicability, including publication bias, p-hacking, and inadequate sample sizes. The small dataset problem is particularly pertinent, as limited data can lead to unstable and unreliable estimates, increasing the likelihood of non-replicable results.

Camerer [12] identified that poor replicability due to frequentist approaches is particularly problematic in social sciences, with replicability rates varying between only 57% and 67% for studies relying on complementary replicability indicators.

Ioannidis [13] published "Why Most Published Research Findings Are False," one of the most downloaded research papers of the decade. The paper demonstrated that for a sample of the most important drugs approved by the FDA in the decade prior to its publication, over half of these research studies could not be replicated. This finding implied that over half of these drugs were either ineffective, a waste of money, or potentially harmful.

From a Bayesian perspective, the issue of poor replicability can be addressed through the incorporation of prior information. Bayesian methods enable researchers to combine prior knowledge with current data to form posterior distributions, resulting in more stable and credible inferences, particularly in the context of small datasets. Priors can be informed by previous studies, expert opinions, or theoretical considerations, providing a way to "borrow strength" from external sources of information. This approach can mitigate the problems associated with small sample sizes by effectively increasing the amount of information available for analysis.

Bayesian hierarchical models are particularly useful for improving replicability with small datasets. These models allow for the pooling of information across different levels or groups within the data, such as different study sites or time points. By sharing information across these groups, hierarchical models can improve the precision of estimates and reduce the impact of random variability. This leads to more robust and generalizable results, enhancing the likelihood that findings will be replicable in future studies.

Another advantage of Bayesian methods in the context of small datasets is the ability to quantify and incorporate uncertainty directly into the analysis. Bayesian inference produces full posterior distributions for parameters, rather than single point estimates. This provides a more comprehensive picture of the uncertainty associated with the estimates and allows for more nuanced decision-making. For instance, instead of relying on a binary significance test, researchers can assess the probability that a parameter lies within a practically significant range, providing a clearer basis for evaluating the robustness of the findings.

Despite these advantages, the success of Bayesian approaches in addressing replicability and small dataset problems depends on the careful selection and validation of priors. Poorly chosen priors can bias the results and undermine the benefits of Bayesian inference. It is crucial to conduct sensitivity analyses to assess how the results vary with different priors, ensuring that the conclusions are robust and not unduly influenced by subjective choices. Transparent reporting of the priors used and their justification is also essential for the credibility and replicability of Bayesian analyses.

The problem of poor replicability and the small dataset problem pose significant challenges in statistical inference. Bayesian methods offer valuable tools for addressing these issues by incorporating prior information, using hierarchical models, and quantifying uncertainty. However, careful consideration and validation of priors are essential to fully realize the benefits of Bayesian approaches. By leveraging these methods, researchers can improve the robustness and replicability of their findings, contributing to more reliable and credible scientific knowledge.

2.2. How Will You Interpret the Results of Your A/B Analysis in Terms that are Relevant to Your Original Problem?

A/B decision models are invoked to discover which of two options is better. These options could involve a product, a policy, an operating environment or something else. In each domain, there are likely to be unique measures of "better". For example, in a business situation, more profit or more revenue are "better". In journalism, more subscribers, or more downloads may be what we say is "better". If we are opining on revenues, then we would like our A/B test statistic to "the change in revenue". However, statistical tests are often not designed to yield such clear-cut metrics, and we have to work around their limitations.

To motivate this section, consider the example graphed in Figure 2.1 that compares a frequentist A/B test to a Bayesian A/B test with noninformative priors. R-code script (included in the Technical Appendix) simulates success/failure data, perform both types of tests, and then graphs the results:

- The data for choice A is generated once with a benchmark conversion rate of 0.40.

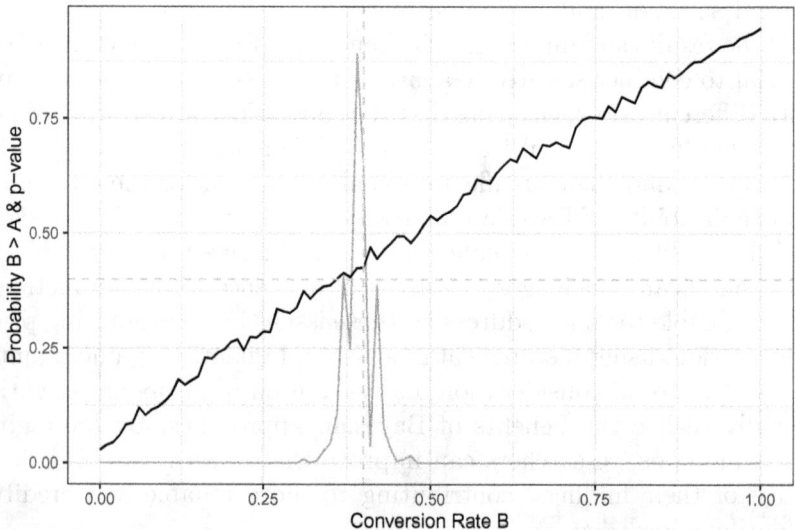

Figure 2.1. Bayesian A/B probabilities (black) versus frequentist p-values (gray).

- The conversion rate for choice B varies from 0 to 1 in increments of 0.01.
- Vectors are initialized to store the p-values from the frequentist tests and the probabilities from the Bayesian tests.
- Priors are uninformative in this case, and the Bayes A/B test only reflects the dataset information.
- The `bayesTest` function from the `bayesAB` R package is used to perform the Bayesian A/B test with a Bernoulli distribution.
- A loop runs over the range of conversion rates for group B. In each iteration, data for group B is generated, and both frequentist and Bayesian tests are conducted. Results from each test are stored in their respective vectors and plotted.

The graph output from this simulation highlights some of the cogent differences in the Bayesian and frequentist approaches to A/B decision models.

Bayesian probabilities that the conversion rate of choice B is higher than that of choice A (fixed at 0.40) steadily rise with the mean of the conversion rate for B. More informative priors will tend to smooth the "Probability $A > B$" line, but otherwise, the general summary of the conclusions will be the same.

Frequentist p-values are probabilities that the conversion rate of choice B is higher than that of choice A which is fixed at 0.40. To the left of the gray peak around 0.40 conversion rate for choice B, the frequentists choose A with p-value ≈ 0; to the right they choose B with p-value ≈ 0. Right around 0.40 there is a considerable amount of uncertainty (high p-values).

The reader should draw several conclusions from this example:

- where choices are clear cut, and ample data is available, frequentist A/B choices are accurate
- when the differences investigated are small — in this example where the conversion rate of B versus 0.40 is less than 0.10 — frequentist tests are easily confused
- p-values may best be interpreted as telling us when and where we cannot make a decision. They provide no information about the benefits of one choice over another
- there is no straightforward method to apply economic information such as costs and profitability to the conclusions of frequentist tests.

Frequentist A/B models generate a test statistic using the available data. By checking the distribution of that test statistic under the null hypothesis, we determine the probability of observing even more extreme statistics, known as the p-value. A small p-value indicates a low probability of such extreme observations, suggesting it's unlikely the statistic was generated under the null hypothesis. Therefore, the data itself is also unlikely to have come from the null hypothesis. The definition of a "small" p-value is up to the statistician's discretion.

Frequentist A/B tests report their decisions in terms of p-values and Neyman-Pearson hypothesis significance testing, which falls short of the standards prescribed in [14]. They present decisions like "$P(A > B)$ is 10%" or "Assuming the null hypothesis that A and B are equal is true, the probability of seeing a result this extreme in A versus B is 3%".

For frequentist A/B testing in particular, interpreting p-values in statistical inference is fraught with challenges and potential misunderstandings. A p-value is the probability of obtaining an observed result, or something more extreme, assuming that the null hypothesis is true. This measure does not provide the probability that the null hypothesis is true or false, nor does it indicate the size or importance of an effect. Misinterpretation of p-values often leads to the erroneous conclusion that a low p-value confirms the alternative hypothesis or that a non-significant p-value proves the null hypothesis. This binary thinking ignores the nuances of statistical evidence and the context of the data.

From a Bayesian perspective, the interpretation of p-values is even more problematic. Bayesian inference focuses on updating the probability of a hypothesis given the data and prior information. In this framework, the p-value is not directly relevant because Bayesian methods do not rely on the null hypothesis significance testing paradigm. Instead, Bayesian analysis uses the posterior probability to assess hypotheses, which provides a more intuitive and informative measure of evidence. The posterior probability directly quantifies the strength of evidence in favor of a hypothesis, given the observed data and prior beliefs.

Bayesian methods offer several advantages over p-value-based frequentist approaches. Firstly, they allow the incorporation of prior knowledge into the analysis, which can lead to more accurate and credible inferences. Priors can be derived from previous studies, expert opinion, or theoretical considerations, and they help to contextualize the data within a broader framework of knowledge. This is particularly useful in cases where data are sparse or noisy, as priors can stabilize estimates and improve the robustness of the conclusions.

Bayesian inference offers a more nuanced understanding of uncertainty compared to traditional methods. Instead of relying on a single p-value to make binary decisions, Bayesian methods generate a full posterior distribution of the parameters of interest. This distribution captures the uncertainty around parameter estimates and allows for more flexible decision-making. For instance, in clinical trials, rather than declaring a treatment effective based solely on a p-value threshold, researchers can evaluate the probability that the treatment effect exceeds a clinically meaningful threshold. This approach provides a richer and more relevant basis for decision-making.

Instead of p-values, Bayesian methods yield direct probabilities on whether A is better than B (and by how much). Rather than point estimates, posterior distributions are parameterized random variables that can be summarized in various ways. Additionally, Bayesian tests are immune to the issue of "peeking" at the data and remain valid regardless of when the test is stopped.

Despite these advantages, Bayesian methods also face challenges, particularly related to the choice and sensitivity of priors. Poorly chosen priors can bias the results, and sensitivity analyses are essential to ensure that conclusions are robust to different prior assumptions. Furthermore, Bayesian methods can be computationally intensive, especially for complex models, though advances in computational techniques and software have made these methods increasingly accessible.

The interpretation of p-values in statistical inference is often problematic and can lead to misleading conclusions. Bayesian methods offer a more informative and nuanced alternative by focusing on posterior probabilities and incorporating prior knowledge. While these methods require careful consideration of priors and computational resources, they provide a richer framework for understanding data and making decisions. As statistical practice continues to evolve, the Bayesian approach is likely to play an increasingly important role in addressing the limitations of p-value-based inference.

2.3. How Do You Choose the Right Power, Confidence, Effect Size and Significance Level for an A/B Test?

The frequentist figures of merit for hypothesis testing are power, confidence and significance level (significance is just a transformation of confidence). The widely used critical value for significance is 0.05, equivalent to 0.95 confidence and proposed by Fisher in the 1920s; for power it is 0.8 proposed by Jacob Cohen in the 1980s. They are frequentist kludges to set critical values for determining the proper sample size. The sample size, in turn, is a surrogate for the amount of information a dataset must contain to provide a reliable decision.

When you specify the significance level of an A/B test, you're making a trade-off between your tolerance for accepting that one experience is better than the other when it really isn't (Type I error or "false positive") versus seeing no statistical difference between the experiences when there actually is a true difference (Type II error or "false negative"). The confidence level is determined before a test is run.

The information in a dataset is captured in its probability distribution, which is functionally a 1-manifold in a two-dimensonal probability space. If there is no information loss in the sample, through multicollinearity or other weaknesses, then three critical zero-dimensional values — significance, power and effect size — can in theory determine the minimum sample size required for a given decision. But a researcher cannot know this in advance; at best they can combine their limited subjective knowledge with a guess about the quality of data collection. You cannot unequivocally map a higher dimensional probability space into a zero-dimensional point.

This leaves frequentists in the same situation as blind men groping an elephant. One touches the trunk, likening it to a snake; another touching the tail describes it as a rope, and the third feels its tusk, concluding that it is a spear. Frequentists set themselves up for decision failures by restricting their protocols to three point statistics, augmented with a host of subjective, unstated assumptions such as Normally distributed data, random selection and so forth.

It gets worse; in situations where data is expensive data collection may be automatically limited by budget constraints. For example, in accounting firms it has been suggested that each audit of a single accounting transaction may cost \$100–200; a budget of \$1,000 would limited collection to 5–10 transactions. In marketing tests it typically is not feasible to run the test long enough to identify the true relative performance of the alternatives, and often the resolution of differences revealed in statistical analysis between the alternatives is too small for any meaningful impact. A "tie" between the two choices often reflects a dataset wasted because of poor analysis, rather than a real tie between the alternatives A and B.

An example can help elucidate the problems of sample size and the frequentists' critical values discussed previously. Assume you are deciding between two online web page formats which received Normally distributed numbers of clicks each month. Web page format B is the current web page format and averages 100 thousand clicks monthly. A is the same page, but modified in a way that will actually attract 10% more clicks. Let $B \sim N(100000, 300000)$ be our "ground truth" — the actual real-world situation that is unknown to the researcher. The researcher "discovers" the truth to some level of resolutions (e.g., within particular confidence limits) by sampling and analyzing monthly outcomes.

As hoped for, the number of incorrect decisions falls with the increase in sample size; the p-values similarly decline. We can interpret the p-value at any sample size as the actual significance level that we would have needed to chose before settling on this dataset sample size. These p-values are larger than Fisher's 0.05 critical value all the way up to a sample size of 10,000. You need a huge sample size before you can actually meet the standards set by Ronald Fisher.

The level of incorrect decisions we see (Figure 2.2) which for sample sizes under 1,000 are 20–30%, may or may not be acceptable, depending on the opportunity cost of each wrong decision. In this problem, if each click-through translates to an average of \$1 of net profit from conversion, then 10% (the increase in CTR with option A) of the 100,000 average CTR would be \$10,000 monthly; if we are making this mistake 20% of the time, that would be \$2,000 expected lost profit due to insufficient sampling. In the next section, I discuss the question of sample size.

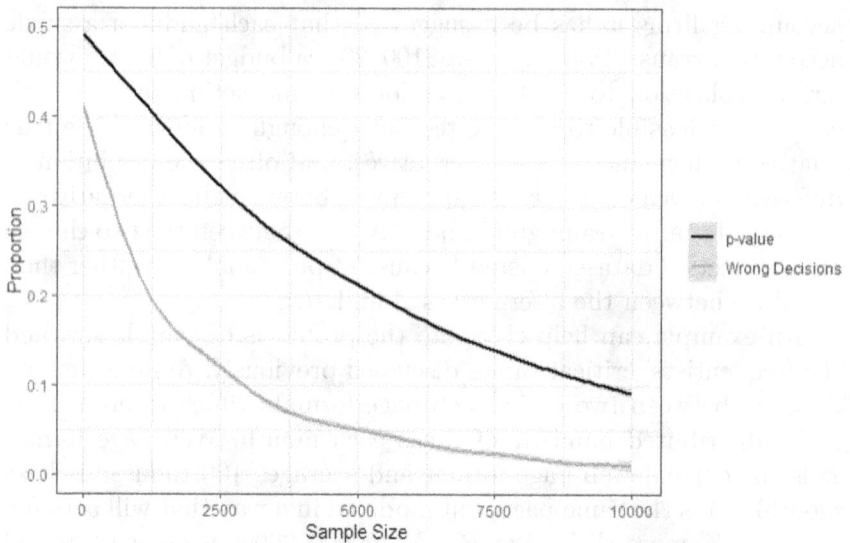

Figure 2.2. Proportion of incorrect decisions for a given p-value.

2.4. What Can You Do if Data Collection is Constrained by Budget, Data Availability or Other Restrictions?

In some natural sciences, e.g., biology, agriculture, chemistry and physics, we can repeat an experiment millions of times a day, generating massive amounts of data. In the extreme, with layer upon layer of complex devices, CERN's LHC detectors generate a staggering petabyte (one million gigabytes) of collision data per second.

In contrast, we often find that in business, medicine, and the social sciences, data may be expensive, of limited quantity, and often uncontrolled, from natural, field or quasi-experimental methods.

Field and quasi-experiments are closely related, but are not appropriate for many social science and business studies; here are the differences:

- Natural experiments rely on an external force (e.g., a government, non-profit, etc.) controlling the randomization treatment assignment and implementation,

- Field experiments require researchers to retain control over randomization and implementation.
- Quasi-experiments occur when treatments are administered as-if randomly (e.g., US Congressional districts where candidates win with slim margins, weather patterns, natural disasters, etc.).

In these cases, individuals or groups are exposed to the experimental and control conditions that are determined by nature or by other factors outside the control of the researcher. The process governing the exposures resembles random assignment, but there are typically not objective methods of determining whether, in fact, data is biased. Nonetheless such methods are necessary when controlled experimentation is difficult to implement or unethical. These are very common problems in political science, population studies, epidemiology and economics [15].

In some fields, there are strong incentives to lie — for example in finance and economics. Pump-and-dump stock manipulation schemes are one such example, where trade volumes may be artificially inflated to spike the price. Crime data may be purposely underreported to make cities seem safer. Frequentist methods are especially vulnerable to such manipulations because they simply summarize their data.

The statistical methods implement in frequentist A/B tests are borrowed from the natural sciences and typically make assumptions of random selection from a well-defined population, independent identically distributed observations, and controlled experimentation. A and B may often be thought of in terms of treatment and control groups. The decisions based on such methods may be called into question when their assumptions are violated, as they are in uncontrolled natural, field or quasi-experimental methods.

Sample size and quality are less of a problem for Bayesian A/B testing because Bayesians *learn* from data rather than using it to generate a one-shot summary statistic. The learning updates the prior distribution to a posterior with different parameter values. Since the prior will contain information not available in the observations, they are relatively less influenced by fake data, and small sample sizes merely cause less of an update to our knowledge.

2.5.   How Much Information is Actually in Your Dataset?

More data is not necessarily better. The amount of information in datasets can vary widely, and this makes any universal measure of significance, power and sample size unreliable and variable.

In particular, economic, financial and accounting data are highly multicollinear. The implication is that you may need to collect, perhaps, ten times as much data to have an equivalent information content to a single observation from a controlled laboratory study, or from a database of customer traffic. The common source of economic, financial and accounting data is accounting transactions. Consider a sale paid with a credit card in a retail store. In accounting systems, that sale is recorded first as a credit to sales revenue and a debit to accounts receivable; simultaneously inventory is credited and cost of goods sold is debited. Later the account receivable is collected and that is recorded as a credit to accounts receivable and a debit to cash. When inventory is replenished, this is recorded as a debit to inventory and credit to accounts payable. And so forth. A single real-world event is turned into a dozen entries to various ledger accounts. And all of those accounts are highly correlated, because you have perhaps a dozen account entries and only one real-world event. Since accounting systems are the basis for all things financial — from accounting reports, to financial analyses to national economic statistics, you see how this mutlicolinearity can be transferred through the whole system. Multicollinearity and weak data are significant issues in statistical modeling that can complicate the interpretation and reliability of results. Multicollinearity occurs when two or more predictor variables in a regression model are highly correlated, making it difficult to discern their individual effects on the response variable. This leads to inflated standard errors, unstable estimates, and reduced statistical power. Weak data, on the other hand, refers to datasets that are either small in size, contain a lot of noise, or have low-quality observations. Such data can lead to imprecise parameter estimates and poor model performance. Addressing these issues is crucial for building robust statistical models, particularly from a Bayesian standpoint.

Bayesian methods offer unique advantages in dealing with multicollinearity and weak data. In a Bayesian framework, prior information can be incorporated into the analysis, which helps stabilize parameter estimates. When faced with multicollinearity, incorporating informative priors can reduce the variance of the posterior estimates, thereby mitigating some of the adverse effects. For instance, if prior knowledge suggests that certain predictors should have small or zero effects, this information can be encoded in the prior distributions, leading to more reliable and interpretable results even in the presence of multicollinearity.

Moreover, Bayesian hierarchical models can be particularly useful when dealing with weak data. These models allow for the pooling of information across different levels of the data, which can be beneficial when individual data points are sparse or noisy. By borrowing strength from related data, hierarchical models can improve parameter estimation and make more robust inferences. This approach is especially valuable in small sample sizes or when dealing with heterogeneous data sources, as it leverages the entire dataset to inform individual parameter estimates.

Another Bayesian technique that helps address the challenges of weak data is the use of regularization priors. Regularization introduces a penalty for large coefficients, which can prevent overfitting and improve the model's generalizability. For example, ridge regression or LASSO (Least Absolute Shrinkage and Selection Operator) can be implemented in a Bayesian context by specifying appropriate priors, such as normal or Laplace distributions. These priors shrink the estimates of less important variables toward zero, effectively reducing the complexity of the model and enhancing its performance on weak data.

Despite these advantages, the success of Bayesian methods in handling multicollinearity and weak data heavily depends on the choice of priors. Poorly chosen priors can lead to misleading inferences and fail to mitigate the issues at hand. Therefore, it is essential to carefully consider and validate the prior distributions used in the analysis. Sensitivity analysis, which involves examining how the results change with different priors, can be a valuable tool in this regard. This helps ensure that the conclusions drawn from the Bayesian model are robust and not unduly influenced by arbitrary prior choices.

2.6.   Have You Found Evidence of Absence, or Just an Absence of Evidence?

Traditional frequentist A/B tests do not inherently differentiate between the absence of evidence and evidence of absence [16,17]. Evidence of absence indicates that the data support the hypothesis that there is no effect (i.e., the two conditions do not differ), whereas, absence of evidence means that the data are inconclusive [18].

Confusing the absence of evidence with evidence of absence is a common error in statistical inference, often leading to incorrect conclusions. Absence of evidence occurs when the data do not provide sufficient information to either support or refute a hypothesis. In contrast, evidence of absence suggests that there is enough information to conclude that an effect or relationship does not exist. Misinterpreting the former as the latter can result in the false assumption that a non-significant result indicates no effect when it may simply reflect insufficient data or power to detect an effect.

When a traditional frequentist A/B test fails to reject the null hypothesis, it is tempting to conclude that the null hypothesis is true. However, a non-significant p-value only indicates that the data are not strong enough to reject the null hypothesis, not that the null hypothesis is proven. This distinction is crucial, emphasizing the importance of considering the study's power, sample size, and effect size when interpreting results. A well-powered study with a large sample size and a small effect size might genuinely provide evidence of absence, but many studies are underpowered and only demonstrate an absence of evidence.

The Bayesian framework offers a more nuanced approach to this issue. Bayesian inference allows researchers to calculate the posterior probability of a hypothesis given the data and prior information. This probability provides a direct measure of the evidence for or against a hypothesis. For example, if the posterior probability that an effect size is practically significant remains low even after considering prior information and the observed data, one can more confidently assert evidence of absence. Conversely, if the posterior distribution is wide and uncertain, it indicates that there is still an absence of evidence, and more data are needed.

The curation of the dataset plays a vital role in addressing the problem of confounding absence of evidence with evidence of absence.

Ensuring high-quality data collection and curation can improve the power and reliability of statistical analyses. This includes careful consideration of sample size, measurement precision, and the inclusion of relevant variables. By enhancing the quality of the dataset, researchers can reduce the likelihood of ambiguous results and make more definitive inferences about the presence or absence of effects.

Additionally, dataset curation should involve a thorough examination of potential confounding variables that could obscure the true relationship between the variables of interest. By accounting for these confounders through appropriate statistical controls or study design, researchers can obtain clearer and more accurate estimates of the effects being studied. This process helps to differentiate between genuine evidence of absence and mere absence of evidence, leading to more robust and reliable conclusions.

This failing — confusing absence of evidence with evidence of absence — is a critical issue in statistical inference that can lead to false conclusions about the presence or absence of effects. Bayesian methods provide a clearer framework for distinguishing between these two concepts by focusing on posterior probabilities. Careful dataset curation, e.g., ensuring adequate sample sizes, measurement accuracy, and controlling for confounders, is essential for making reliable inferences. By addressing these issues, researchers can avoid common pitfalls and improve the validity of their statistical conclusions.

2.7. How Will You Handle Sequential Tests and Streaming Data Updates?

Many A/B decision applications require periodic or continuous updates of data. Traditional A/B testing is not well-suited for making decisions based on sequential or continuous updates. Companies like Amazon aim to update product listings on their website continuously, implying a need to stream feedback into their A/B decision models. Real-time updating is common in online retail, stock portfolio management, inventory management, and various other fields. However, frequentist A/B models are poorly suited for handling streaming updates.

Traditional A/B testing does not account for time, even though most applications require its consideration. For instance, many datasets are streamed in real-time, such as stock prices, breaking news, and financial reports. However, traditional A/B testing treats data as a single batch. There are frequentist workarounds, like decision trees and multi-armed bandits, which incorporate sequence and infer time from data, but these methods are unique to traditional A/B testing. With Neyman–Pearson tests, data cannot be tested sequentially without necessitating a correction for multiple comparisons, which depends on the sampling plan. This issue is discussed in [19–22].

In contrast, Bayesian A/B testing is inherently suited to scenarios where time is a critical factor, as it "learns" from each new data point and piece of information. This makes Bayesian methods perfectly natural for making decisions that involve time, which applies to nearly every important decision.

Sequential testing in frequentist statistical inference refers to the practice of repeatedly analyzing data as it is collected, with the potential to stop the study early if results appear significant. While this approach can be appealing for its efficiency and ethical considerations, it introduces substantial risks to the validity of the results. The primary issue with sequential testing is that it inflates the Type I error rate, or the probability of falsely rejecting the null hypothesis. Each additional test conducted increases the chance of finding a significant result purely by chance, undermining the integrity of the statistical conclusions.

In a frequentist framework, the problem of inflated Type I error rates due to sequential testing is often addressed through methods such as the Bonferroni correction, which adjusts the significance threshold to account for multiple comparisons. However, these corrections can be overly conservative, reducing the power of the tests and making it more difficult to detect true effects. Another approach is the use of predefined stopping rules, where the criteria for stopping the experiment are determined before any data are collected. While these methods help to control error rates, they still face challenges in maintaining a balance between controlling Type I errors and retaining sufficient power to detect true effects.

From a Bayesian viewpoint, sequential testing is naturally accommodated, as Bayesian inference continuously updates the probability of a hypothesis as new data becomes available. In Bayesian analysis, there is no strict requirement to fix the sample size or the number of looks at the data in advance. Instead, the focus is on the posterior distribution, which reflects the current state of knowledge given the data and prior information. This continuous updating process inherently adjusts for the accumulation of evidence over time, providing a more flexible and intuitive framework for sequential testing.

Bayesian decision-theoretic approaches offer additional tools to manage sequential testing. For example, Bayesian methods can incorporate predefined stopping rules based on the posterior probability of a hypothesis or the expected utility of continuing versus stopping the experiment. These stopping rules are grounded in the Bayesian framework, which naturally adjusts for multiple looks at the data. This allows researchers to make informed decisions about when to stop the experiment based on the evolving evidence, without inflating the risk of false positives.

Despite these advantages, Bayesian sequential testing is not without its challenges. The choice of prior distributions can significantly impact the results, and poorly chosen priors can introduce biases. It is essential to conduct sensitivity analyses to ensure that the conclusions are robust to different prior assumptions. Additionally, Bayesian methods can be computationally intensive, particularly for complex models or large datasets. Advances in computational techniques and software are making Bayesian methods more accessible, but careful consideration of computational resources and methodologies remains crucial.

Sequential testing poses significant challenges in frequentist statistical inference due to the risk of inflated Type I error rates. While frequentist methods such as corrections for multiple comparisons and predefined stopping rules can mitigate these issues, they often come with trade-offs in statistical power. Bayesian approaches offer a more flexible and intuitive framework for sequential testing by continuously updating the posterior distribution as new data is collected. This allows for more adaptive and informed decision-making, although careful consideration of priors and computational demands is necessary. By leveraging the strengths of Bayesian methods, researchers can better manage the complexities of sequential testing and make more reliable and robust inferences.

2.8.   Am I "Peeking"?

In statistical inference, "peeking" refers to the practice of repeatedly analyzing data at various stages of an experiment to determine whether to stop early. This practice can lead to inflated Type I error rates, as the probability of finding a statistically significant result by chance increases with each additional analysis. In a frequentist framework, this issue is particularly problematic because it violates the assumption that the test is conducted only once, thereby compromising the integrity of the p-value and potentially leading to false positives.

From a Bayesian perspective, the concept of peeking is addressed differently. Bayesian inference naturally accommodates continuous data monitoring, as it updates the posterior distribution as new data becomes available. This allows Bayesian methods to potentially handle interim analyses without the same risk of inflated error rates seen in frequentist approaches. However, this does not imply that peeking is entirely without consequences in Bayesian analysis. If not properly managed, repeated looks at the data can still introduce biases and affect the credibility of the conclusions.

To mitigate the issues associated with peeking, Bayesian practitioners often use predefined stopping rules or Bayesian decision-theoretic approaches. These methods specify in advance the criteria under which the experiment will be stopped, thereby controlling the overall error rates and ensuring that the inference remains valid. For instance, Bayesian stopping rules might be based on the probability that a certain parameter exceeds a clinically meaningful threshold or that the posterior distribution reaches a specific level of certainty. By adhering to these predefined rules, researchers can maintain the integrity of their analyses while benefiting from the flexibility of Bayesian updating.

Moreover, Bayesian approaches can incorporate prior information that reflects the costs and benefits of different decision outcomes, further refining the stopping criteria. This incorporation of decision theory allows for a more holistic approach to experiment design and data analysis, balancing the risk of false positives against the potential for early discovery of meaningful effects. For example, in clinical trials, Bayesian methods can weigh the ethical considerations of continuing a trial when early data suggest a significant benefit or harm, thus making more informed and ethical decisions.

Despite these advantages, the implementation of Bayesian stopping rules and the management of peeking require careful planning and transparent reporting. Researchers need to clearly articulate their stopping rules and ensure that these rules are strictly followed to avoid post-hoc decision-making, which can reintroduce bias. Additionally, sensitivity analyses should be conducted to assess how different stopping rules and priors impact the results, providing a robust framework for evaluating the credibility of the findings.

While the Bayesian framework offers a more flexible and theoretically sound approach to handling peeking in statistical inference, it is not without its challenges. Properly managed, Bayesian methods can mitigate the risks associated with repeated data analysis, maintaining the validity of the inferences drawn. However, this requires careful planning, transparent reporting, and rigorous sensitivity analyses to ensure that the benefits of Bayesian updating are fully realized without compromising the integrity of the results.

Bayesian methods provide several benefits over frequentist methods in the context of A/B tests — namely in interpretability. Instead of p-values, you get direct probabilities on whether A is better than B (and by how much). Instead of point estimates, your posterior distributions are parametrized random variables which can be summarized in any number of ways. Bayesian tests are also immune to "peeking" and are thus valid whenever a test is stopped.

Monitoring A/B tests until a significant result is achieved might seem practical, but it adversely impacts the effective statistical significance of the test. This greatly increases the likelihood of false positives and makes confidence intervals untrustworthy. To avoid these pitfalls, determine an adequate length of time for the test to run before initiating it. While it's acceptable to monitor the test for proper implementation, avoid drawing conclusions or stopping the test prematurely.

Similarly, stopping tests prematurely based on early results can be misleading. Early performance fluctuations are often just random variations and not indicative of long-term trends. To prevent these errors, allow the test to run until a sufficient number of observations have been reached, adhering to the planned duration and sample size to ensure accurate and reliable results.

2.9. Is all the Information Relevant to My Decision Captured in My Dataset?

What is easy to measure is not necessarily what we need to measure to answer the problem we are analyzing. In many cases, we often tend to obtain datasets for A/B testing that are convenient — survey data from tools like SurveyMonkey, published financial data, and data scraped from websites. This is what is available, and it is suitable for some questions, but not for others. But accurately capturing the essence of your decision extends beyond simply measuring what is easily quantifiable. While these sources provide convenient access to data, they may not always align with the specific needs of the analysis, posing challenges in addressing the core questions effectively.

The frequentist approach traditionally relies on data formatted homogeneously, as seen in a typical Excel spreadsheet of observations. This perspective assumes that all necessary information for decision-making is contained within these observations, a presumption that often does not hold true in real-world scenarios. By contrast, Bayesian inference recognizes that data sets represent only a fraction of the available knowledge about a problem. This approach allows for the integration of a broader spectrum of information, including historical data, expert opinions, and subjective beliefs, through the use of prior distributions.

Frequentist A/B tests primarily utilize the data at hand without incorporating prior knowledge, which may lead to conclusions that neglect valuable contextual information such as historical trends or expert insights. This limitation can be particularly significant in scenarios involving small sample sizes or high data variability, where external knowledge could provide crucial stabilization and depth to the analysis.

Bayesian methods, however, excel in integrating this external information through prior distributions, which encapsulate beliefs about the parameters before data observation. These priors are updated with new data to form posterior distributions via Bayes' theorem, offering a more comprehensive view that includes both prior knowledge and new evidence. This capability not only enriches the understanding of the parameters involved but also enhances decision-making processes by allowing for the calculation of expected values and other statistics crucial for risk assessments and cost-benefit analyses.

Moreover, Bayesian analysis is dynamically adaptive, continuously updating its conclusions as new data becomes available, which is especially advantageous in rapidly evolving fields. However, the effective implementation of Bayesian methods necessitates careful selection and justification of priors, along with rigorous sensitivity analyses to ensure the robustness of the findings against various assumptions.

In a Bayesian framework, the initial step involves mathematically encapsulating prior beliefs by selecting a distribution that reflects where parameters are believed to lie. As new data is collected, it is integrated with the priors to update the posterior distribution, providing a real-time probabilistic view of the parameters. This approach mitigates common pitfalls in frequentist A/B testing, such as the increased risk of false positives from repeated hypothesis testing without the requisite data accumulation.

Furthermore, Bayesian inference can address issues related to novelty effects in testing scenarios. For instance, changes to a well-established website feature may initially disrupt returning visitors' engagement, temporarily affecting the performance of a new offer. By segmenting visitors into new and returning groups and comparing conversion rates, one can discern whether performance discrepancies are due to novelty effects or genuine inferiority of the offer. This segmentation helps in understanding long-term impacts and adjusting strategies accordingly.

Bayesian methods offer a more nuanced and flexible framework that accommodates the full spectrum of information available, making it possible to draw more informed, robust, and contextually relevant conclusions.

2.10. How Should I Treat the Population Parameter for Each Variant?

In statistics, the term "parameter" differs from its broader mathematical usage, specifically referring to a quantity that characterizes a statistical population. Parameters such as the mean or standard deviation serve to summarize or describe aspects of the population.

When a population adheres to a known and defined distribution, such as the normal distribution, it can be completely characterized by a small set of parameters. These parameters not only describe the population but also define a probability distribution, which is essential for drawing samples from that population.

The relationship between a "parameter" and a "statistic" mirrors the relationship between a population and a sample. A parameter represents a true value computed from the entire population, like the population mean, while a statistic is an estimate of this parameter derived from a sample, such as the sample mean. Therefore, a "statistical parameter" is more precisely known as a population parameter, reflecting its role in defining and describing the entire population.

Frequentist and Bayesian methodologies differ in their treatment of population parameters. Frequentists consider these parameters as unknown constants — fixed but not directly observable quantities that must be estimated from data. In contrast, Bayesians treat each parameter as a random variable with a probability distribution, a distinction that significantly influences the type of inferences and calculations possible within each framework.

Frequentist approaches typically utilize point estimates and confidence intervals for analysis. A point estimate represents the best single guess of an unknown parameter, while a confidence interval provides a probable range for this parameter, predicated on the assumption of repeated data collection processes. This methodology does not integrate prior knowledge and focuses solely on the information presented by the current dataset, hence it does not provide a probabilistic assessment of the parameter itself. As a result, frequentist statistics are somewhat limited in expressing the uncertainty and variability of parameter estimates in probabilistic terms.

Conversely, Bayesian methods provide a more dynamic and comprehensive framework for statistical inference by viewing parameters as random variables. This perspective enables the incorporation of prior distributions, reflecting pre-existing knowledge or beliefs about these parameters before new data are observed. Upon the acquisition of new data, Bayesian statisticians update these priors to form posterior distributions via Bayes' theorem, encapsulating both prior beliefs and new evidence.

The capability to model parameters as random variables and generate their probability distributions offers substantial advantages. It allows for a direct examination of these distributions, enabling a more detailed understanding of uncertainty. For instance, Bayesian analysts can compute credible intervals which indicate the probability bounds within which parameters exist, directly derived from the posterior distribution. This method extends beyond simple point estimates or intervals, providing a full spectrum of statistical understanding.

Furthermore, Bayesian methods facilitate the computation of expected values and other summary statistics from the parameter distributions, proving invaluable in decision-making scenarios where such statistics can guide cost-benefit analyses and risk assessments. In contexts like clinical trials, Bayesian statistics can estimate the probability that a new treatment outperforms an existing standard, including the anticipated magnitude of effect — thereby supporting more informed decision-making.

Additionally, the Bayesian framework allows for the continuous updating of beliefs as new data becomes available, ensuring that statistical inferences remain current and comprehensive. This contrasts with frequentist methods, which often require reanalysis of the entire dataset or complex sequential testing corrections to integrate new information, a process that can be both cumbersome and counterintuitive.

This distinction in the way that Bayesians treat population parameters — viewing them as fixed constants versus probabilistic variables — enables Bayesian methods to offer a richer, more adaptable framework for statistical analysis. By capitalizing on the strengths of Bayesian inference, researchers can achieve deeper insights into parameter uncertainty and variability, leading to more robust and informed statistical conclusions.

2.11. Have You Changed the Streaming Data Allocations During the Testing Period?

A/B testing may be performed on datasets that are updated by streaming data, offering flexibility to implement minor adjustments during the testing process. A common modification involves altering the distribution of traffic between test experiences, a change that can temporarily influence the accuracy of results until the data stabilizes.

For instance, consider an A/B test where 80% of the traffic is initially directed to Experience A (the control) and 20% to Experience B. If, during the testing period, the traffic allocation is adjusted to an equal 50% split and later shifted entirely to 100% for Experience B, it's important to understand the impact on user assignment.

In this scenario, when the allocation is changed to 100% for Experience B, users originally placed in Experience A continue in their initial assignment. This alteration affects only those newly entering the test.

Should there be a need to significantly modify the traffic percentages or alter how visitors are distributed between experiences, it is advisable to start a new testing activity or duplicate the existing one, followed by editing the traffic allocation percentages accordingly.

When adjustments are made to the traffic shares during the test, several days may be required for the data to normalize, particularly if there is a high occurrence of returning visitors.

Furthermore, if an A/B test originally has a 50/50 traffic split which is then adjusted to 80/20, then initial results can appear distorted. This is especially true if the average time to conversion is lengthy, taking several hours or days. For example, in a scenario where the traffic split shifts from 50% to 80%, and the average conversion time is two days, only visitors from the initial 50% might convert on the first day of the new split. This could misleadingly suggest a decline in the conversion rate, but normalization is expected once the newly adjusted 80% of visitors have also reached the two-day conversion point.

2.12.   Does Your A/B Modeling Strategy Reflect Your Organization's Strategy?

The effectiveness of A/B testing can be greatly influenced by how well the statistical models align with the organization's strategic objectives. For both frequentist and Bayesian approaches, the choice and alignment of objective functions, or loss functions, with business goals are vital.

Frequentist methods typically focus on minimizing error rates, such as Type I and Type II errors, or maximizing statistical power. In the context of A/B testing, these methods often prioritize easily measurable metrics like click-through rates (CTRs) or conversion rates. While these can yield quick results, they may not align with broader organizational goals such as maximizing revenue, customer lifetime value, or brand loyalty. For instance, an A/B test might show that a variant increases CTR significantly; however, if these clicks do not lead to meaningful engagements or purchases, the test fails to advance the organization's overarching objectives. This misalignment can divert resources toward strategies that do not foster desired business outcomes.

On the other hand, Bayesian approaches provide a more adaptable framework, enabling the incorporation of organizational goals directly into the analysis. By treating parameters as random variables with associated probability distributions, Bayesian methods can integrate prior knowledge and business objectives into the model. This integration is achieved through a loss function that quantifies the true costs or benefits of different outcomes, taking into account both direct and indirect effects on organizational objectives. For example, a Bayesian A/B test could employ a loss function that more heavily penalizes scenarios with low revenue conversions compared to those with low CTRs, thereby ensuring that strategic priorities like revenue maximization are central to the decision-making process.

However, the effectiveness of Bayesian methods hinges on the accurate definition and integration of the organization's goals into the prior distributions and loss functions. Inaccurate priors or poorly aligned loss functions can lead to suboptimal decisions. It is imperative for researchers to deeply understand the business context and collaborate closely with stakeholders to establish appropriate priors and loss functions that genuinely reflect the organization's aims. Conducting sensitivity analyses is also crucial to ensure that conclusions are robust against variations in prior assumptions and loss function specifications.

In both the frequentist and Bayesian frameworks, the key to effectively supporting organizational objectives is through meticulous selection and alignment of metrics, priors, and loss functions with the actual business goals. This demands a collaborative approach, where researchers engage with business leaders to fully grasp the broader strategic objectives and incorporate these insights into the statistical models used for decision-making. By aligning the objective functions used in A/B testing with the organization's goals, researchers can deliver more relevant and actionable insights, ultimately driving superior business outcomes.

In practice, marketers may be tempted to use metrics like CTR in the upper funnel to quickly gather a sufficient number of test conversions due to their high traffic and low variance. However, it's crucial to evaluate whether CTR genuinely reflects the intended business objective. For example, an offer generating a high CTR might attract visitors less likely to purchase, or the nature of the offer, such as a discount, might reduce overall revenue. This is illustrated by a scenario where a skiing offer generates a higher CTR than a cycling offer, yet the latter results in higher average spending per visitor, leading to greater overall revenue. Therefore, an A/B test favoring the skiing offer based on CTR alone does not align with the primary business objective of revenue maximization.

To circumvent these issues, it is essential to carefully monitor and understand the actual impact of different metrics on business outcomes, ideally employing a metric that more closely aligns with your business goal.

Chapter 3

Understanding Prior Distributions

3.1.   What is a "Probability"?

Consider tossing a fair coin. "Fair" means that the coin will come up heads 50% of the time and tails 50% of the time. At least *in the long run*. No one would be surprised if you tossed the coin twice and it came up heads both times. This "in the long run" caveat suggests a particular strategy for defining probability. For just a few coin tosses, we wouldn't be surprised at almost any outcome. But as we do more and more tosses, we expect the total proportion of heads to come closer to 50%. This notion of what we mean by probability is called "frequentism" and defines probabilities as the relative frequency with which a particular outcome shows up after an infinite number of trials.

Frequentists mistakenly assume this perspective is "objective" because their probabilities are, ostensibly, a feature of the coin being tossed. But infinity is a *big* number — indeed, no one ever reaches it. In practice the percentages of heads in a fair coin toss will differ from 50%, though by a decreasing amount with more trials. But it *is* different, and that error grows exponentially when problems exhibit more choices for outcomes — for example with six-sided dice, or networks with hundreds of thousands of nodes. Furthermore in many important research areas — e.g., medicine, sociology, business, economics — we have limited numbers of observations. In these areas, frequentism fails with great regularity. One famous paper titled "Why Most Published Research Findings Are False" [13] found that more than half of drug trials that were approved by the FDA based on frequentist arguments turned out to be false.

3.2. Epistemic Probability

This book advocates an alternative approach to understanding probability, focusing on *epistemic* probability. Epistemic probability measures what we know rather than appealing to an abstract and unachievable infinite number of trials. In this framework, we assign a *credence* — essentially a degree of belief — to various propositions we are considering. These credences function similarly to probabilities, ranging from 0 to 100%, and the total set of credences for all possible outcomes of a specific event must sum to 100%.

Statisticians have formalized this procedure under the label of Bayesian statistics, named after the Reverend Thomas Bayes, an 18th-century Presbyterian minister and amateur mathematician. Bayes developed a formula that outlines how to adjust our credences based on new information. Bayesian statistics importantly incorporate our beliefs and prior knowledge into decision-making processes.

Bayesianism also acknowledges the significance of data, such as results from tossing a coin. Data informs our understanding of the situation, aiding decisions about, for example, whether a coin is truly "fair". If initially you assign a 50% credence to the coin landing on heads, but then observe five consecutive tails, you might adjust your credence to perhaps 20% or 30%. This epistemic view of probabilities is inherently subjective, recognizing that different individuals, based on their unique states of knowledge, may assign different credences to the same event. However, this variability is acceptable as long as everyone agrees to update their credences consistently when new evidence emerges. Thomas Bayes provided a key rule — Bayes' theorem — for guiding how individuals should revise their beliefs in light of new data.

Bayesian inference takes the observed data, distilled into a *likelihood function*, and multiplies it by the *prior*, which has distilled unobserved information into a single distribution, to generate the *posterior* distribution with the same functional form as the prior, so that the process can be used over-and-over in a learning process as the last decision's posterior becomes the next decision's prior (Figure 3.1).

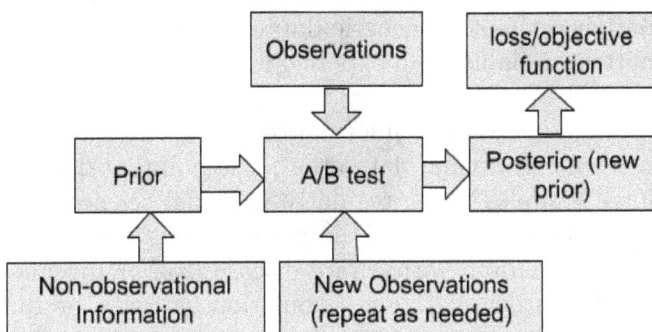

Figure 3.1. Schematic of bayesian A/B decision making.

In Bayesian probability theory, if a likelihood function (i.e., the probability function that summarizes the data for a given model) is in the same probability distribution family as the prior probability distribution, then the prior and posterior are called *conjugate distributions* with respect to that likelihood function. A conjugate prior is an algebraic convenience, giving a closed-form expression for the posterior; otherwise, numerical integration might be necessary. Furthermore, conjugate priors intuitively show how a likelihood function updates a prior distribution.

The parameters of the prior and posterior are called *hyperparameters* to distinguish them from the parameters of the underlying model (the likelihood function). The form of the conjugate prior can generally be determined by inspection of the probability density or probability mass function of a distribution. A typical characteristic of conjugate priors is that the dimensionality of the hyperparameters is one greater than that of the parameters of the original distribution.

It is these hyperparameters that are used to capture the knowledge we have about our decision that are external to our data set. There are several common sources for this non-dataset information:

- Our subjective beliefs about the context in which the decision and statistical inference will occur.
- Results of prior studies, or done by other parties concerning this or similar decisions; in this case we say the prior is being used to "pool" information.
- The posterior distribution for Bayesian inference with the same data but from an earlier time.

- Information available in generic databases and
- Information available from expert opinions.

Often this non-database information is the only way to make an inference using a particular dataset, especially if that data set is very small. Small samples occur regularly in social sciences for various reasons. Sometimes the size of the population is extremely limited, for example in children with a rare disease [23], or juvenile females charged with murder [24]. The population can also be difficult to recruit and prone to drop-out because they are homeless, institutionalized, or playing truant [25]. Factors such as costs [26] and ethical constraints [27] may also make efforts to obtain a larger sample quite difficult (or impossible).

Prior distributions can avoid inadmissible estimates and convergence issues in Bayesian estimation. Several simulation studies including [28,29] have shown that typical frequentist estimation methods like maximum likelihood estimation suffer from these problems. In sum, Bayesian estimation with informative priors results in meaningful output, even with small samples, and can increase the precision of the result when prior information is available.

3.3.   The Secret Lives of Information

Datasets and observed data are not the sole sources of information for your A/B test. In fact, for any problem faced by individuals or firms, there is a wealth of information available, often surpassing what can be gathered in a dataset. This information can be invaluable, even though it is not necessarily in a matrix format to ease its use in computer software or mathematical analysis. It may exist in textual form, in the perceived outcomes of previous experience, or in a "common sense" understanding of the problem space. Traditional A/B testing often overlooks these rich sources of information, because it doesn't know how to use it. Bayesian A/B analysis stands out in being able to quantify and incorporate these sources of information into a rigorous mathematical framework, providing the analyst with unique advantages over their frequentist counterparts.

These external sources of information, not derived from the observed data under analysis, are collectively known as "prior" information. In Bayesian statistics, they are established prior to the inference process. During inference, the observed data is distilled into a likelihood function. This function, as its name suggests, represents the likelihood of our observed dataset occurring if our statistical model of the real world and its set of parameter values is accurate. In other words, it is the joint probability of observed data under the assumption that our statistical model accurately reflects the real world.

It's important to note that even in traditional frequentist A/B modeling, there exist prior beliefs. This is a common thread that runs through both Bayesian and frequentist statistics, highlighting the shared understanding and principles in statistical modeling. But frequentists' priors are opaque, and motives in incorporating subjective choices are not as transparent as they should be. Using frequentist approaches, we impose our own prior beliefs in the model's structure, the choice of the dataset to gather, and the features that we measure, e.g., gas mileage, number of cylinders, and so forth in a car dataset. These prior beliefs, in traditional statistical inference powerfully influence the final decision. In attempts to justify *ad hoc* methods used to incorporate this "prior" information, frequentists have imposed a profusion of convoluted structures on their underlying models. This trend increases the setup problems associated with traditionalist A/B testing, creates problems in the efficient use of observations in the dataset, and confounds interpretation. Without an explicit role for this unobserved information, decisions and inferences will be brittle, non-robust, and weak. Furthermore, frequentist models have difficulty learning from subsequent observations, data, successes, and failures.

Bayesians have given external, non-dataset information explicit roles in their inference. They are given a particular structure in Bayesian A/B tests — the prior distribution — which has a particular role to play in decision making. This process also yields decisions in the form of probability distributions, and these are ultimately easier to interpret. This chapter explains how Bayesians incorporate unobserved information into their decision processes.

3.4.   Lies, Damned Lies, and Statistics

Mark Twain reputedly was quoted saying, "There are three kinds of lies: lies, damned lies, and statistics". Certainly similar claims have been made about the reliability of decisions based on A/B tests and the way that they can be manipulated to support any prejudice or desired outcome. Marketers, for example, may quip that that A/B tests are the drunk holding up the light post (i.e., the choice the marketer would prefer to see). Perhaps this is not totally fair, but it highlights the brittleness and lack of robustness in traditional A/B testing. If I am looking for a significance of 5% (a 0.05 p-value) in the outcome, I can run 20 A/B tests, and pick the one that supports the choice I want to see; or I can keep collecting data until I get a $1/20=0.05$ p-value. This is called "peeking" and is, unfortunately, a too-common practice in commercial or political applications of A/B testing.

Traditional A/B testing lacks robustness partly due to poor practice and protocols. Most software for hypothesis testing (A/B testing is a variation on hypothesis testing) assumes data are Normally distributed without regard to the underlying real-world processes generating the data. So practitioners will just use the software and parameters that are easily available to crank out their results, and assert that because a computer generated these results are valid. They adopt a version of cargo cultism that appropriates the trappings of technology, assuming that is enough for legitimacy.

The biggest complaint lobbed toward Bayesian statistics is that it is complex, and indeed, there is truth in this concern. But because it delivers more interpretable and accurate results using the same data, that added complexity is justified in nearly every instance. Bayesian complexity arises from the need for the researcher to specify two pivotal aspects of the real world that the traditionalists ignore; the Bayesian must understand:

- how the real world generates the data from which the researcher's models will learn,
- how to incorporate the wealth of information available for any problem that lies outside the dataset.

Bayesian models are more complex than frequentist models because they can consider more information about the researcher's problem and are able to use that to generate more accurate and interpretable decisions. Bayesian complexity comes with the promise of creating robust, reliable, and interpretable decisions. Bayesian models are inherently more powerful because they incorporate more information than traditional models (from the priors). Bayesian models are inherently more easily interpreted because they generate a parameterized distribution (the posterior) that provides reliable estimates, and allows valuation with a loss function. Contrast this with traditional frequentist A/B models that generate a zero-dimensional point estimate — the p-value — rather than the complete two-dimensional distribution of probabilities in the posterior.

All of this is great theory, but let's be pragmatic. Which would you rather present to your boss?:

- I chose **A** because the p-value $= 0.07639$, or
- If we choose **A**, we can expect to generate \$10,000 more profit for the company and the risk of a loss greater than \$5000 is 15%. If you provide me with potential future scenarios (e.g., best, expected, worst) I can provide you with expected financial outcomes, and risk profiles. And so forth.

3.5. Formal Strategies for Choosing Priors

In Bayesian analysis, specifying a prior distribution is essential to account for uncertainty about model parameters. Unlike frequentist methods, where the prior is implicit (maximum likelihood estimation) or specified via a penalty (penalized maximum likelihood estimation), Bayesian approaches require explicit prior specification.

In the current book, we will restrict ourself to conjugate priors, i.e., priors that have identical functional form to their posteriors. Thus choice of a prior will depend on the nature of the data and likelihood function, e.g., is it count data, or continuous, or binary. We can assume that the precision (inverse variance) of the prior can be viewed as an implied sample size. For large sample sizes, the posterior mean converges to the maximum likelihood estimate, making the prior less critical. In these cases, we may choose parameters that imply a large variance. But where the data is restricted to small sample sizes, a prior must be explicitly specified.

There are two formal approaches to prior selection which have been suggested in the literature: (1) weakly informative forms; and (2) empirical Bayes methods. It is important to understand that there are no truly uninformative priors. A prior distribution that looks uninformative (i.e., "flat") in one coordinate system can be informative in another — this is a simple consequence of the rule for transformation of probability densities. Furthermore, suggestions for uninformative priors are likely to be improper, i.e., they do not sum to 1.

The best known noninformative prior is given by a proposal by Harold Jeffreys [30] for a prior that is invariant against transformation of the coordinate system of the parameters. The Jeffreys prior is constructed from the expected Fisher information using the likelihood:

$$p(\theta) \propto \sqrt{\det I(\text{Fisher}(\theta))}$$

For the Beta-Binomial model the Jeffreys prior corresponds to the U-shaped $Beta(\frac{1}{2}, \frac{1}{2})$.

For the normal-normal model it corresponds to the flat improper prior $p(\mu) = 1$ and for the IG-normal model the Jeffreys prior is the improper prior $p(\sigma^2) = \frac{1}{\sigma^2}$.

The main problem with Jeffreys priors is their functional form: they are not conjugate, i.e., prior and posterior are different distributions. Furthermore, many Jeffreys priors are not probability distributions, as they do not sum to 1.

Reference priors were developed by [31] as an alternative to Jeffreys priors. They maximize the mutual information between the data and the parameter, thereby minimizing the influence of the prior. In univariate settings the reference priors are identical to Jeffreys priors, but can be extended to multivariate settings. They suffer the same shortcomings as Jeffreys priors.

Another class of prior constructing strategies is termed "empirical Bayes". In empirical Bayes strategies the data analysis specifies a family of prior distribution and then the data at hand are used to find an optimal choice for the hyper-parameters (hence the name "empirical"). Thus, the hyper-parameters are not specified but themselves estimated.

Given data D, a likelihood $p(D|\theta)$ for a model with parameters θ, and a prior $p(\theta|\lambda)$ for θ with hyperparameter λ, the marginal likelihood now depends on λ:

$$p(D|\lambda) = \int_\theta p(D|\theta)p(\theta|\lambda)d\theta$$

We can use maximum (marginal) likelihood to find optimal values of λ given the data. This type of empirical Bayes is also known as "Type II maximum likelihood".

An alternative way to estimate hyperparameters is by minimizing the empirical risk. In Bayesian estimation, the posterior mean of the parameter of interest is obtained by linear shrinkage:

$$\hat{\theta}_{\text{shrink}} = E(\theta|D) = \lambda\theta_0 + (1 - \lambda)\hat{\theta}_{\text{ML}}$$

Here,

- θ_0 is the prior mean (also called the "target").
- $\hat{\theta}_{\text{ML}}$ is the maximum likelihood estimate (MLE).
- λ is the shrinkage intensity, determined by the ratio of the prior and posterior concentration parameters (k_0 and k_1).
- The resulting point "shrinkage" estimate $\hat{\theta}_{\text{shrink}}$ is a convex combination of θ_0 and $\hat{\theta}_{\text{ML}}$.

The hyperparameter in this setting is k_0 (linked to the precision of the prior) or equivalently the shrinkage intensity λ. An optimal value for λ can be obtained by minimizing the mean squared error of the estimator $\hat{\theta}_{\text{shrink}}$.

By construction, the target θ_0 has low or even zero variance but non-vanishing and potentially large bias.

3.6.   Practical Strategies for Choosing Priors

In my opinion, neither the Jeffreys nor the empirical Bayes approaches provide particularly useful guidance in practice. Problems arise because the prior information we have for a typical situation comes in multiple forms — prior studies, subjective beliefs, current news, diverse opinions from friends and coworkers, and so forth. These may need to be combined and weighted, and not every source will wield commensurate authority. Thus, the process of curating what we already know becomes complex and subjective in itself.

In practice, constructing a conjugate Bayesian prior with appropriate hyperparameters involves a process where you incorporate existing knowledge about a parameter of interest into a prior distribution. This is typically done when you want to update this prior with data to get a posterior distribution. The conjugate prior is chosen because it makes the mathematical process simpler — specifically, it ensures that the prior and posterior distributions are in the same family, which facilitates easier analytical solutions. There are three steps to constructing a conjugate Bayesian prior.

Step 1: *Define the likelihood for the dataset and choose the conjugate prior*

Identify the likelihood function based on the data type and the distribution it follows. For example, if your data is binary, you might use a Bernoulli distribution; for count data, a Poisson distribution might be appropriate; for continuous data, consider distributions like the Normal distribution. Once you have identified the likelihood, choose a prior distribution that is conjugate to this likelihood. I delve more deeply into these choices later in this chapter.

Step 2: *Set the conjugate prior's hyperparameters based on existing knowledge*

This is the most critical step where you use existing knowledge to set the hyperparameters of the prior distribution. Consider these points:

- **Quantitative knowledge:** If you have specific data or estimates from previous studies, use these to estimate the parameters. For example, if you know the average and variability of a parameter from previous data, use these to set the mean and variance of your prior.
- **Qualitative knowledge:** For more subjective assessments (like expert opinions), translate these into parameters. For instance, how confident is the expert, and how does this influence the spread or scale of the prior?

Step 3: *Reality check*

Finally, update this prior with actual data to get the posterior distribution. The conjugate properties ensure, for example, that if you start with a Beta prior and use a Bernoulli likelihood, your posterior will also be a Beta distribution, with updated parameters based on the data. This systematic approach helps in integrating both objective data and subjective judgements into a coherent Bayesian framework. In this step you are able to assess whether or not your prior is dominating the data, or conversely, whether the data (likelihood function) ignores the prior. You may wish to alter your prior hyperparameters to more accurately reflect the certainty or authority that you would like to assign to your prior knowledge.

Prior information can be obtained from a myriad of sources, objective and subjective. If the sources are objective, then we may refer the process of Bayesian revision as "pooling". For example, in meta-studies, we may pool the conclusions from many studies, conducted in different environments, under different assumptions, with different subjects and with different levels of authority and accuracy. Additionally, researchers with a knowledge of the subject area will have a wealth of qualitative and quantitative knowledge about the problem and the data.

The Prior Assessment Matrix (PAM) is a tool that allows each source of prior information to be explicitly documented, and consolidated into statistics that will define the parameterization of the prior. These priors for our Bayesian analysis of any dataset will be, for our purposes, summarized in location μ and dispersion σ — the estimates of the population mean and standard deviation. These are familiar statistics that have simple, easily understood and communicated interpretations. I have chosen standard deviation rather than variance, because it has the same units as the mean, and thus is intuitive. The PAM is a matrix with columns:

- Source.
- Estimated mean.
- Estimated standard deviation.
- Authority (high = 3, medium = 2, low = 1).

There is no preferred method for consolidation of these information sources, and researchers can adopt their own methods, so long as there is a conservatism that favors a higher value for standard deviation. A higher value means that the data in the likelihood will have more influence on the posterior.

Source	Estimated mean stars	Estimated standard deviation	Authority
My beliefs	5	8	Medium = 2
IMDB	4	11	High = 3
RT	4	8	High = 3
	—	—	—
Averages	4.5	9	NA

Consider an example with a dataset that polled a particular movie audience to see how they liked the movie "American Fiction". We might compute sample mean and standard deviation from the star ratings in user reviews at *IMDB* and *Rotten Tomatoes*. We thus have two sources to plug into our PAM, and assuming each site has similar authority, we can average the values for sample mean and standard deviation and choose prior parameters that give us those values. Estimates $| \mu = 4.25 | \sigma \approx 9 | - |$.

In the above example, I have encoded the authority as a weight multiplier. Thus

$$\mu = \frac{\sum_{\text{row}=1}^{3} \text{Est.Mean} \times \text{authority}}{\sum \text{authority}}$$

and

$$\sigma = \frac{\sum_{\text{row}=1}^{3} \text{Est.SD} \times \text{authority}}{\sum \text{authority}}.$$

If the researcher is uncertain whether there is useful or authoritative information that can be incorporated in the prior distribution, then construction of the prior can be more cavalier and compensated for with an artificially large variance.

For a Normal–Normal conjugate family, this would mean that the prior mean was the pooled sample averages, and the prior standard deviation was the pooled sample standard deviations. For other priors like the Gamma and Beta, we would similarly solve for their prior hyperparameters. For example if value X is Beta distributed, then

$E[X] = \frac{\alpha}{\alpha+\beta}$, and $\text{Var}[X] = \frac{\alpha\beta}{(\alpha+\beta)^2(\alpha+\beta+1)}$. These two equations in two unknowns can be solved for the prior hyperparameter values α and β.

3.7. The World in Six Distributions

We commonly use one of six different probability distributions (listed in Table 3.1) to model randomness in the real world. Each lends itself to specific types of situations, specific ways of measuring the data from that situation, and in the process of learning from trying out one or another management policies to implement our models and generate new data (e.g., investing to generate ROI data).

Why do these specific distributions appear so often. There may be specific processes; for example addition tends to produce Normally distributed random variables. There may be a single real-world process, but if we measure two different aspects of the process we get two different distributions. For example, we might want to model the rate of calls coming into a computer help desk. We could either use the number of calls per hour (Poisson distribution) or the time between each incoming call (Exponential distribution). The results

Table 3.1. Parameters of the six distributions.

Likelihood	Where used	Parameters	Prior	Prior hyperparameters
Bernoulli	Successes	p	Beta	α successes and β failures
Poisson	Counts	λ (rate)	Gamma	α occurrences in β intervals
Exponential	Waiting times	$\frac{1}{\lambda}$ (delay)	Gamma	α observations that sum to β
Normal	Additive	μ and σ^2	Normal-inverse gamma	ν, μ_0, α, β
LogNormal	Multiplicative	μ and σ^2	Normal-inverse gamma	ν, μ_0, α, β
Uniform	Values limited to a maximum	$U[0, \theta]$	Pareto	k with maximum value x_m

will be the same, but the measurement, modeling, and conjugate families are different.

Bayesian A/B tests commonly use six different distributions to model specific sets of processes that generate observed data. This is another area in which Bayesians tend to be more transparent. Traditional A/B testing can be coerced to consider the various ways that observed data may be measured and generated. But, for convenience traditional tests often choose a Normal distribution and justify this using the Central Limit Theorem. Most software is set up to compute and report statistics for the Normal distribution, so assuming a Normal distribution minimizes software setup, and supports reported statistics that are widely understood. The problem with this is that many, if not most important real world processes are decidedly not Normal. Consider asset prices, like stocks, bonds, and houses; such price series tend to be normal. Or telephone calls to a call center, responses to an advertisement, or customers arriving at a service desk; these tend to be exponentially distributed over time, or Poisson distributed in blocks of time, e.g., per hour. Normal assumptions in these cases will lead to incorrect decisions.

We say that we have "conjugate" priors when the data can be summarized in a particular form for the likelihood function that when multiplied by the "conjugate" prior, results in a posterior with the same form as the prior. For example, data summarized as a Poisson likelihood, when multiplied by a Gamma distribution results in a posterior that is also a Gamma distribution. A list of conjugate priors for our six distributions appears in Table 3.1.

As shown in Table 3.1, the most commonly encountered datasets fall into one of six categories, and each of these categories defines a specific form for the likelihood function — Bernoulli, Poisson, Exponential, Normal, LogNormal and Uniform. In turn, each of these has a conjugate prior that when updated with the data encapsulated in the likelihood function will yield a posterior distribution with the same form as the prior — i.e., which is conjugate (Table 3.2). Here are the six common distributions that model processes in the real world, where we would use A/B testing to guide our decisions.

(1) **Bernoulli: Successes and failures** are best modeled with a *Bernoulli* distribution. The Bernoulli distribution is appropriate where your observed data is well-modeled by 1s and 0s, according

Table 3.2. Constructing priors from estimated mean and standard deviation.

Prior	μ and σ of the prior	Hyperparameters as function of μ and σ
Pareto	$\dfrac{\alpha * x_m^2}{(\alpha-1)^2(\alpha-2)} = \sigma^2$ and $\dfrac{\alpha * x_m}{(\alpha-1)} = \mu$	$\alpha = \sigma^2 - \dfrac{\sqrt{\mu^2 * \sigma^2 + \sigma^4}}{\sigma^2}$ and $x_m = \mu^2 + \sigma^2 + \dfrac{\sqrt{\mu^2 * \sigma^2 + \sigma^4}}{\mu}$
Gamma	$\dfrac{\alpha}{\beta^2} = \sigma^2$ and $\dfrac{\alpha}{\beta} = \mu$	$\alpha = \dfrac{\mu^2}{\sigma^2}$ and $\beta = \dfrac{\mu}{\sigma^2}$
Normal-inverse-gamma	μ and σ	$\mu = \mu$, $\beta = \sigma^2$, $\lambda = 1$, and $\alpha = 2$,
Beta	$\dfrac{\alpha * \beta}{(\alpha+\beta)^2(\alpha+\beta+1)} = \sigma^2$ and $\dfrac{\alpha}{\alpha+\beta} = \mu$	$\alpha = \dfrac{\mu^2 - \mu^3 - \mu * \sigma^2}{\sigma^2}$ and $\beta = \dfrac{(\mu-1) * (\mu^2 - \mu + \sigma^2)}{\sigma^2}$

to a specific probability p of a 1 occurring, e.g., a "success". For example:

- Click-through-rate/conversions for a web page.
- Heads and tails in a coin flip.

The conjugate prior for a Bernoulli likelihood function is the $Beta(\alpha, \beta)$ distribution. Given our prior estimates μ and σ from the PAM we can solve the following system of two equations:

$$\frac{\alpha * \beta}{(\alpha + \beta)^2(\alpha + \beta + 1)} = \sigma^2$$

$$\frac{\alpha}{\alpha + \beta} = \mu$$

This yields prior hyperparameters α and β:

$$\alpha = \frac{\mu^2 - \mu^3 - \mu * \sigma^2}{\sigma^2}$$

$$\beta = \frac{(\mu - 1) * (\mu^2 - \mu + \sigma^2)}{\sigma^2}$$

(2) **Normal: Averages, totals, and amounts generated through addition and subtraction** yield data that is best modeled with a *Normal* distribution.

The Normal distribution is appropriate where your observed data is the result of addition and subtraction. For example:

- Account balances such as total sales, that are the sum of numerous individual sales transactions.
- Average sales across 100 stores, where the average is the sum of sales divided by the number of stores.

The Normal distribution is arguably the most important distribution in statistics. It is the default distribution we turn to if we lack further information about the source or form of the data. Because of this, statisticians have explored analysis with Normal data to a further extent than with any other distribution. In Bayesian revision with Normal data used here, we make the assumption that both the mean and standard deviation parameters of the Normal are unknown, and that data are exchangeable and identically distributed.

The conjugate prior for a Normal likelihood function where we assume that both the mean and standard deviation are unkown is the normal-inverse-gamma distribution. Suppose $x \,|\, \sigma^2, \mu, \lambda \sim \mathrm{N}(\mu, \sigma^2/\lambda)$ has a normal distribution with mean μ and variance σ^2/λ where $\sigma^2 \,|\, \alpha, \beta \sim \Gamma^{-1}(\alpha, \beta)$ has an inverse-gamma distribution. Then (x, σ^2) has a normal-inverse-gamma distribution. Start our construction of the prior by noting that λ is simply an inverse scale factor for the variance, and for prior construction we can fix this to $\lambda = 1$. We can replace the distribution of σ^2 with its mean $\sigma^2 = \frac{\beta}{\alpha - 1}$ then let $\alpha = 2$ and $\beta = \sqrt{\sigma}$. That is quite a number of simplifications, but all involving the dispersion σ. Since prior dispersion involves subjective assessments of certainty, authority or accuracy, this simplification will have no effect on our ability to construct a prior distribution that reflects what we know.

Given our prior estimates μ and σ from the PAM, plus our assumptions about the hyperparameters that define the prior on the variance we have:

$$\lambda = 1$$

$$\alpha = 2$$

$$\mu = E[x] = \mu$$

$\beta = \frac{\beta}{(\alpha-1)\lambda} = \text{Var}[x] = \sigma^2$ and $\sqrt{\beta} = \sqrt{\text{Var}[x]} = \sigma$ is the standard deviation

(3) **LogNormal: Return on investment, stock price time series, and ammounts generated from multiplication and division** are best modeled by a *lognormal* distribution.

The lognormal distribution is similar to the Normal distribution but is appropriate where your observed data is the result of multiplication and division. For example:

- Stock prices tend to rise or fall as a percentage of the current price of a share of stock; this is true for the prices of many types of asset.
- Return on investment is a normalized increment in the price of an asset, and tends to follow a lognormal distribution.

When data is lognormally distributed, it is typical to transform it so that it becomes Normally distributed. If the random variable X is lognormally distributed, then $Y = \ln(X)$ has a Normal distribution. In this case, we would natural logarithmically transform the data first, and then proceed with Bayesian revision for a Normally distributed random variable.

(4) **Poisson: Counts**, such as the number of times a web server is accessed per minute, the number of telephone calls arriving at a call center per day, and other "count" statistics are best modeled by a *Poisson* distribution.

The Poisson distribution is appropriate where your observed data is measured over time, and reports the average number of events per time interval. Poisson distribution to count data (number of events) has been used in many situations, for example:

- customers arriving at a store,
- number of insurance losses or claims occurring in each period,
- risk of large earthquakes,
- decays in each time interval in a radioactive sample,
- soldiers killed by horse-kicks each year in each corps in the 19th-century Prussian cavalry.
- yeast cells used when brewing Guinness beer,
- goals in sports involving two competing teams,

- deaths per year in a given age group,
- jumps in a stock's price in each time interval,
- times a web server is accessed per minute,
- cells infected at a given multiplicity of infections,
- bacteria in a certain amount of liquid,
- photons arriving at a telescope,
- the molar mass distribution of a living polymerization,
- the number of mutations on a strand of DNA per unit length, telephone calls arriving at a call center.

The conjugate prior for a Poisson likelihood function is the $Gamma(\alpha, \beta)$ distribution. Given our prior estimates μ and σ from the PAM we can solve the following system of two equations:

$$\frac{\alpha}{\beta^2} = \sigma^2$$

$$\frac{\alpha}{\beta} = \mu$$

This yields prior hyperparameters α and β:

$$\alpha = \frac{\mu^2}{\sigma^2}$$

$$\beta = \frac{\mu}{\sigma^2}$$

(5) **Exponential: Waiting and delay times**, such as the amount of time spent on a web page or in a store before leaving and the amount of time a customer service desk may expect to spend handling each request, are best modeled using an *exponential* distribution.

The Exponential distribution is appropriate where your observed data is measured over time and reports the average time between events. It is like the Poisson distribution but reflects a different measurement process where the time between events is measured, rather than the number of events in a time interval. Exponential distributions are used in situations similar to Poisson distributions, but where waiting times rather than counts are measured. For example:

- The rate of decay of radioactive materials.
- Time spent on a web page.

The conjugate prior for an Exponential likelihood function is the *Gamma*(α, β) distribution. Given our prior estimates μ and σ from the PAM we can solve the following system of two equations:

$$\frac{\alpha}{\beta^2} = \sigma^2$$

$$\frac{\alpha}{\beta} = \mu$$

This yields prior hyperparameters α and β:

$$\alpha = \frac{\mu^2}{\sigma^2}$$

$$\beta = \frac{\mu}{\sigma^2}$$

(6) **Uniform: Where observed data is subject to a strict maximum**, at a randomly chosen time, the amount of data will follow a *uniform* distribution.

Uniform distributions are used in estimating, for example:

- The maximum total inventory size from individually numbered snapshots.
- The probability of choosing a particular item in a "random" sample.
- The probability that your gas tank will contain a certain amount of gas at a randomly chosen time.

Uniform distributions are relatively easy to generate with common software, and so may often be the focus of fraud testing. For example:

- Benford tests often assume that manipulated data is more uniform than actual data and use simple statistics to highlight potentially fraudulent data or theft.
- Data manipulation in Harvard professor Francesca Gino's publications was identified when researchers noticed a graph showing the age of cars in an inventory was uniform (in fact it should have a long right tail since cars wear out over time).

The conjugate prior for a Uniform likelihood function is the *Pareto*(α, x_m) distribution. Given our prior estimates μ and σ from the PAM, and noting that the Pareto's mean exists only for $\alpha > 1$ and variance only for $\alpha > 2$ we can solve the following system of two equations:

$$\frac{\alpha * x_m^2}{(\alpha - 1)^2(\alpha - 2)} = \sigma^2$$

$$\frac{\alpha * x_m}{(\alpha - 1)} = \mu$$

This yields prior hyperparameters α and x_m

$$\alpha = \sigma^2 - \frac{\sqrt{\mu^2 * \sigma^2 + \sigma^4}}{\sigma^2}$$

$$x_m = \mu^2 + \sigma^2 + \frac{\sqrt{\mu^2 * \sigma^2 + \sigma^4}}{\mu}$$

3.8. Weak and Strong Priors

Priors can have a significant effect on the shape of the posterior, even with adequate data impacted in the likelihood function. You should have no problems with any reasonable choice of mean and standard deviation defining the prior, and if you leave the model setup open to inspection by any party that might be interested in scrutinizing your analysis. But it is possible to bias the posterior with extreme values of the prior.

Choosing priors correctly is important. Here are some ways to leverage objective/diffuse (assigning equal probability to all values) priors:

- Beta(1, 1).
- Gamma(eps, eps) ~ Gamma(0.00005, 0.00005) will be effectively diffuse.
- InvGamma(eps, eps) ~ InvGamma(0.00005, 0.00005) will be effectively diffuse.
- Pareto(eps, eps) ~ Pareto(0.005, 0.005) will be effectively diffuse.

Keep in mind that the Prior Plots for `bayesTest`'s run with diffuse priors may not plot correctly as they will not be truncated as they approach infinity.

The following example shows how a change in priors can have a significant effect on the posterior distribution. In this example I use Poisson distributed data, where choice A has a mean $\lambda = 45$ and choice B has a mean $\lambda = 46$. The prior is Gamma with parameters α and β and prior mean $\mu = \frac{\alpha}{\beta}$ and prior variance $\sigma^2 = \frac{\alpha}{\beta^2}$. I compare three sets of priors:

(1) $\mu = 1$ and $\sigma^2 = 1$.
(2) $\mu = 100$ billion and $\sigma^2 = 10$ quadrillion.
(3) $\mu = 10$ trillion and $\sigma^2 = 10$ quadrillion.

The graph in Figure 3.2 shows the results for a Gamma prior with $\mu = 1$ and $\sigma^2 = 1$.

The graph in Figure 3.3 shows the results for a Gamma prior with $\mu = 100$ billion and $\sigma^2 = 10$ quadrillion.

The graph in Figure 3.4 shows the results for a Gamma prior with $\mu = 10$ trillion and $\sigma^2 = 10$ quadrillion.

In the last two examples, the assumptions underlying the prior overpower the data in the likelihood, pushing the posterior means up to ~1000 and ~10 million, respectively. In a practical setting, one would question a model setup where the data had a mean of ~45, but we were then compelled to assume that the prior mean was 10 quadrillion with a variance of 10 quadrillion. Certainly this screams

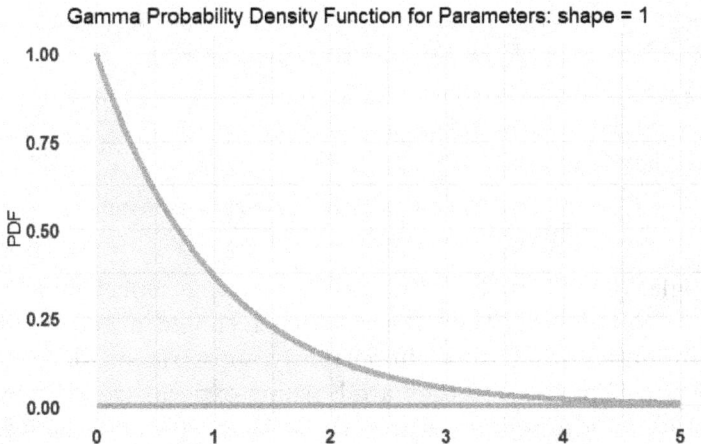

Figure 3.2. Small hyperparameter values.

Gamma Probability Density Function for Parameters: shape = 1e+06

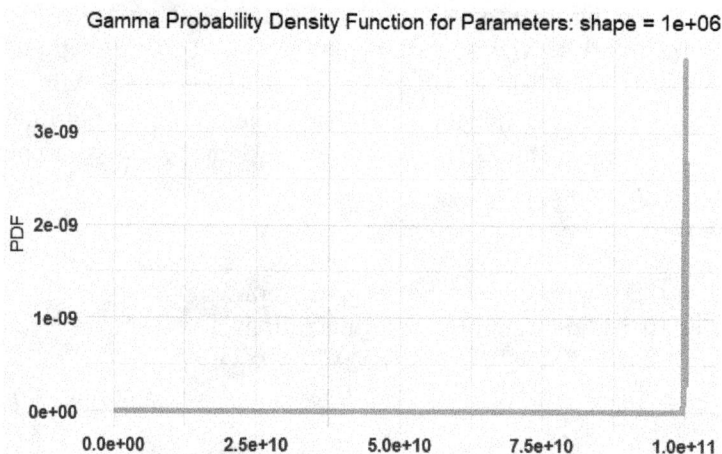

Figure 3.3. Large hyperparameter values.

Gamma Probability Density Function for Parameters: shape = 1e+10

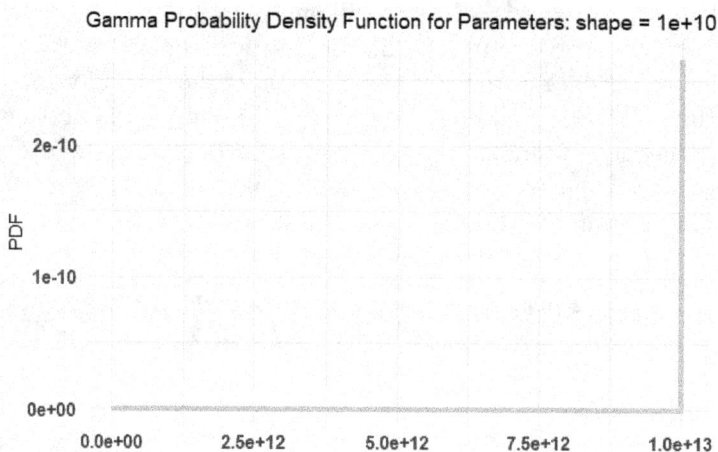

Figure 3.4. Extremely large hyperparameter values.

"manipulation". If, indeed, there is considerable uncertainty at the start of A/B testing, then the researcher should choose a diffuse, non-informative prior. More importantly, most real-world problems are not one-shot choices; they are learning experiences that process sequential updates to the researcher's decision with updated tranches of observations. Frequentist A/B tests need to set up a new on-shot A/B test with every new tranche of data. Bayesians can set the new prior equal to the previous inference's posterior in a true learning model fashion. Additionally, where there are multiple sources of observations, these can be pooled in the same fashion.

3.9. Applications

The initial three chapters of this book have systematically presented the rationale for employing Bayesian A/B decision analysis, alongside methodologies for structuring such analyses. Subsequent chapters will examine posterior distributions and discuss the practical application of Bayesian A/B decision models in specific scenarios, using case studies to address common challenges encountered in the field.

Chapter 4

Posterior Distributions

4.1. Which Posterior Should I Use?

Bayesian A/B analysis yields three posteriors: the one for choice A, the one for choice B, and the one for the difference $B-A$. Posterior analysis may use one or more of these depending on the problem. For example, assume you have two posterior distributions for sales, representing two alternatives to promote the product. Assume one of your objectives is to ensure that there is no stock out of inventory; your current inventory level is 800 units, and your objective is to stay in stock. You are likely interested in the risk (i.e., the tail area of the posterior) of more than 1000 units of sales (Figure 4.1).

The objectives you choose are essential in evaluating the posteriors. If, for example, your decision on whether choice A is better than choice B is based on the expected number of unit sales, then choice A with an expected 600 unit sales is better than choice B at 400 units. But if your objective is to minimize stockouts, then the area under the posterior tail is your figure of merit. We can find from the quantile function that the posterior for B has 2% of its density in the tail above 800 unit sales; the posterior for A has 15%, or about seven times the probability of stocking out as choice A:

- If your objective is the highest expected sales, then A is the preferred choice.
- If your objective is the lowest risk of stockout, then B is the preferred choice.

65

Density Functions of Two Posterior Distributions

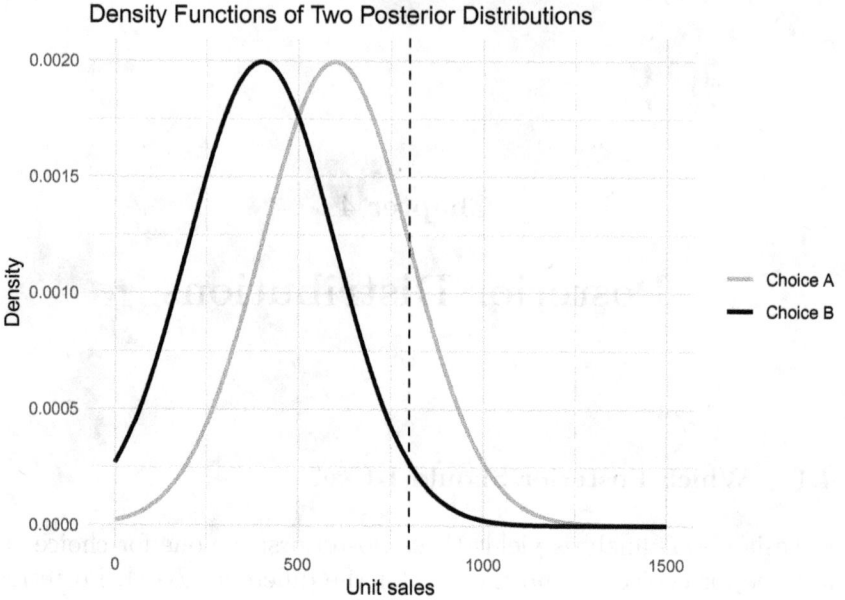

Figure 4.1. Posterior distributions.

Consider an additional layer of information provided by the firm's cost structure. Assume that this product has a setup cost of $1,000 and a marginal cost for each additional unit sold of $1. Let the sales price be $5 per unit, and the firm's objective is to maximize profit. Then our profit function is

$$\text{profit} = \text{units} \times \left[5 - 1 - \frac{\$1000}{\text{units}} \right] = \$4 \times \text{units} - \$1000.$$

Now let's consider a profit maximizing objective for our A/B decision. In this situation, neither A nor B is a clear winner across all sales levels. Rather, B runs losses up until sales of 250 units; between 250 and 500 units, B is the clear winner, and above 500 units, A is the clear winner (Figure 4.2).

4.2. The Economic Value of a Decision

In mathematical optimization and decision theory, a loss function (or cost function) maps an event or values of one or more variables

Expected Total Profit for Number of Units Sold

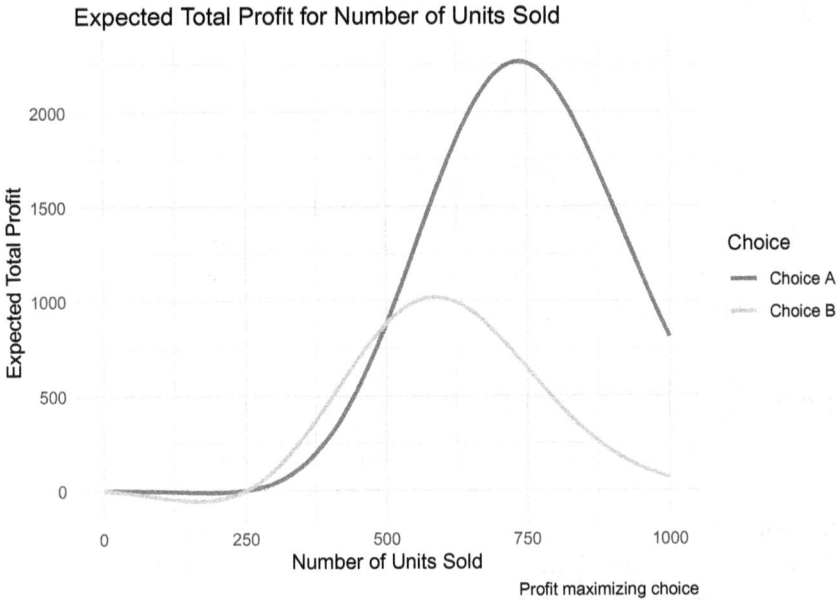

Figure 4.2. For profit application of posteriors.

to a real number, representing the "cost" associated with the event. The objective of an optimization problem is to minimize this loss function. An objective function can either be a loss function or its opposite (referred to as a reward function, profit function, utility function, fitness function, etc.), in which case the goal is to maximize it. The loss function may include terms from various levels of the hierarchy.

In statistics, a loss function is typically used for parameter estimation, representing some function of the difference between estimated and true values for a data instance. This concept, as old as Laplace, was reintroduced in statistics by Abraham Wald in the mid-20th century. In economics, it often represents economic cost or regret. In classification, it represents the penalty for incorrect classification of an example. In actuarial science, it models benefits paid over premiums in an insurance context, a practice popularized by Harald Cramér in the 1920s. In optimal control, the loss represents the penalty for failing to achieve a desired value. In financial risk management, the function is mapped to a monetary loss.

4.3.   Commonly Used Loss Functions

4.3.1   *Regret*

Leonard J. Savage argued that when using non-Bayesian methods such as minimax, the loss function should be based on the concept of regret. Regret is defined as the difference between the consequences of the best decision that could have been made, had the underlying circumstances been known, and the decision that was actually made before those circumstances were known.

4.3.2   *Quadratic loss function*

The quadratic loss function is commonly used in least squares techniques. This function is mathematically tractable due to its symmetry: an error above the target results in the same loss as an error of the same magnitude below the target. If the target is t, the quadratic loss function is expressed as

$$\lambda(x) = C(t - x)^2$$

for some constant C. The value of the constant does not affect the decision and can be normalized by setting it to 1. This function is also known as the squared error loss (SEL).

Many common statistical methods, including t-tests, regression models, and experimental design, use least squares methods grounded in linear regression theory, based on the quadratic loss function. The quadratic loss function is also employed in linear-quadratic optimal control problems. In these problems, even in the absence of uncertainty, achieving the desired values of all target variables may not be possible. Often, loss is expressed as a quadratic form in the deviations of variables of interest from their desired values; this approach is tractable because it results in linear first-order conditions. In stochastic control, the expected value of the quadratic form is used. Due to its squared nature, the quadratic loss function assigns more importance to outliers than to the true data, so alternatives like the Huber, Log-Cosh, and SMAE losses are used when the data contains many large outliers.

4.3.3   0–1 *Loss function*

In statistics and decision theory, a frequently used loss function is
the 0–1 loss function:

$$L(\hat{y}, y) = [\hat{y} \neq y]$$

using Iverson bracket notation. This function evaluates to 1 when
$\hat{y} \neq y$ and 0 otherwise.

4.4.   Risk Management

Bayesian methods excel in risk assessment, providing a full distribution of outcomes that allows for exploration of specific scenarios,
such as losses greater than a deductible amount or the risk of default
on credit loans.

  In essence, risk is the potential for adverse outcomes, encom-
passing the uncertainty regarding the effects or implications of an
activity on elements valued by humans, such as health, well-being,
wealth, property, or the environment, with a particular emphasis
on negative consequences. Numerous definitions of risk exist, but
the international standard definition is "the effect of uncertainty on
objectives".

  The perception, assessment, and management of risk, as well as its
descriptions and definitions, vary across diverse fields, including business, economics, environmental science, finance, information technology, health, insurance, safety, and security. Each of these practice
areas approaches the understanding, assessment methods, and management strategies of risk differently, reflecting the unique challenges
and priorities inherent to each domain.

4.5.   Pooling Information

Combining Bayesian posterior distributions involves synthesizing
information from multiple sources into a single posterior distribution
that represents updated beliefs after observing data. This process

is fundamental in Bayesian statistics, enabling the incorporation of prior knowledge with observed evidence.

Combining Bayesian posterior distributions typically occurs through Bayesian updating, where each posterior distribution serves as a prior for the next update. This iterative process involves multiplying the prior distribution by the likelihood function, which represents how compatible the observed data are with each possible parameter value, and then normalizing to obtain the updated posterior distribution.

In practice, combining multiple posterior distributions can occur in various scenarios, such as:

- **Sequential Bayesian Updating:** When new data becomes available over time, Bayesian posterior distributions can be updated sequentially, with each new observation refining the estimates obtained from previous data.
- **Hierarchical Modeling:** Bayesian hierarchical models allow for the combination of information from different levels of a hierarchy. Each level contributes to the overall inference process, with information flowing from lower-level observations to higher-level parameters.
- **Meta-Analysis:** In meta-analysis, Bayesian methods can combine results from multiple studies to produce an overall estimate of an effect size or parameter of interest. Each study's posterior distribution contributes to the synthesis of evidence across studies.
- **Mixture Modeling:** Bayesian mixture models combine multiple component distributions to model complex data patterns. Each component distribution represents a distinct source of variation, and the mixture model provides a flexible framework for capturing heterogeneity in the data.

Overall, combining Bayesian posterior distributions enables the integration of diverse sources of information, leading to more robust and informed inference in a wide range of applications, from scientific research to decision-making in industry and policy.

4.6. Case Study: The Movie Critics

Linda and Karen had been the best of friends since their college days. Living in a quiet suburb of Chicago, their weekly movie nights were a sacred tradition, a little slice of joy amidst their routine lives as housewives. This week, however, presented a unique dilemma with two appealing options at the local cinema: "Madame Web" and "Mother of the Bride".

On a sunny Thursday afternoon, they sat in Linda's cozy kitchen, each armed with their iPads, scrolling through movie reviews while sipping on chamomile tea. Linda, always the more tech-savvy of the pair, had several tabs open with reviews from top critics.

"So, what are they saying about 'Madame Web'?" Karen asked, peering over her glasses. She had a soft spot for superhero movies, a guilty pleasure that made her feel a bit more youthful.

Linda, tapping on her iPad, read aloud, "It seems 'Madame Web' is getting mixed reviews. One critic says it's a 'bold take on lesser-known comic book characters with a thrilling plot and stunning visual effects.' However, another critic argues that it's 'a cluttered narrative that fails to capture the depth of its central character.'"

Karen pursed her lips thoughtfully. "And 'Mother of the Bride'?"

Switching tabs, Linda found the information. "This one seems to be more positively received. Here's a review that says, 'A heart-warming comedy that brilliantly portrays familial relationships, with standout performances from the lead cast.' Seems like it's more than just a simple comedy."

As Karen cleared the table, Linda's mind raced with a sudden burst of inspiration. Despite their inclination to watch "Mother of the Bride," Linda couldn't help but think about a more analytical approach to their movie dilemma. Being the tech-savvy one, she was always on the lookout for ways to incorporate more data-driven decision-making into her daily life. "Karen, wait", Linda exclaimed, a spark of excitement in her voice. "What if we used a Bayesian A/B test to decide?"

Karen, already half-way to the sink with a stack of plates, turned around with a bemused smile. "A Baye-what now?"

"Bayesian A/B testing", Linda explained quickly, her fingers already flying over her tablet's screen. "It's a statistical method people use to compare two versions of a web page, or in our case, two movies, to see which one is better based on actual data."

Karen chuckled and sat back down, her curiosity piqued. "Alright, Professor Linda, show me this high-tech wizardry."

Linda quickly found a database of movie reviews and pulled up the latest ratings from IMDB and Rotten Tomatoes for both "Madame Web" and "Mother of the Bride". She then began to set up a Bayesian A/B testing model on her tablet using the R statistical software. The idea was to compare the probabilities that one movie would be more enjoyable than the other based on user ratings.

First, she inputted the ratings from IMDB into her model. She calculated the likelihood of each movie's average rating being the better choice based on their score distributions. "See, Karen, we use the reviews as data points to inform our test."

Karen nodded, slowly getting the hang of it. "So it's like the reviews are voting for the best movie?"

"Exactly!" Linda replied, pleased with Karen's analogy. Next, she moved on to the Rotten Tomatoes scores, inputting the fresh percentages into her model. Each source of reviews updated the prior probabilities, refining the results of the Bayesian test. Her screen flashed the probability that each movie would be a better pick, adjusting dynamically with each new data point.

Linda and Karen watched the probabilities shift slightly with each input. The final output was intriguing: the probabilities suggested a near even split, but slightly favoring "Mother of the Bride." However, Linda had an additional trick; she adjusted the model to account for a scenario where the choice would "equally disappoint", aiming to find a balance in potential enjoyment.

"The Bayesian test suggests 'Mother of the Bride' is the safer bet, but it's so close that picking 'Madame Web' would only be marginally less satisfying," Linda concluded, showing the final probabilities to Karen.

Karen laughed. "So after all that, we're almost back to where we started?"

Linda grinned, a little sheepish. "Pretty much. But at least we used science to get there!"

How did Linda do it? She was able to find the critics' and user review datasets for both films on the IMDB and Rotten Tomatoes websites. Keeping it simple, she chose a Poisson-Gamma conjugate family for Bayesian revision. The average star ratings from datasets she pulled form the Internet were:

Source	Madame web	Mother of the bride
A/B choice	A	B
Prior beliefs	4	4
IMDB	2	3
Rotten tomatoes	3	2

Linda's challenge was to perform Bayesian A/B tests separately for IMDB and then Rotten Tomatoes datasets of critics' and user reviews, and then combine the posteriors to make her final choice.

Figure 4.3 shows the final posterior distributions for Choice A (Madame Web) and Choice B (Mother of the Bride) from which Linda drew her conclusions. At the moment those twin Poisson peaks popped up on Linda's screen, both women laughed, the room filling

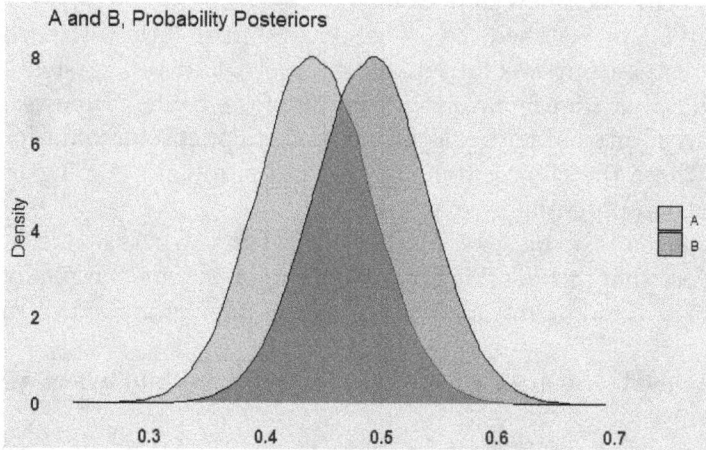

Figure 4.3. Mother of the bride (choice B, dark gray) is the winner.

with the warmth of their shared amusement. They soon gathered their things, ready to leave for their movie night, content with their choice and the fun they had in making it.

As they drove to the cinema, the conversation never waned, ranging from the reliability of movie reviews to the unexpected joy of learning something new. From then on, their movie nights were never just about the film; they were about the experience and the unbreakable bond they shared, a perfect blend of spontaneity and methodical Bayesian decision-making.

Chapter 5

Financial Application of Bayesian A/B Decision Models

This and the following chapters demonstrate real-world applications of Bayesian A/B decision models through a series of case studies. Some of these case studies use real-world data, some use synthetic data, and a few present hybrid synthetic and real data. Each case study presents a common business decision problem, chooses the data that would be needed to make a decision, and implements the Bayesian A/B test. I have tried to keep the presentations simple and concise but with the key information needed to make a decision. This will provide the reader with the necessary guidance to incorporate the Bayesian A/B decision models used here into their own bespoke analyses. The R language code for this and subsequent implementations is available in the Technical Appendix at the end of this book.

Bayesian A/B decision models have become an invaluable tool in the field of financial applications, offering a structured approach to decision-making under uncertainty. These models utilize Bayesian statistics to provide more nuanced insights compared to traditional frequentist methods, allowing for more informed and flexible decision-making processes. However, while Bayesian A/B decision models have numerous applications and advantages, they also come with their own set of challenges and pitfalls that must be carefully managed. This narrative explores the various applications of these models in finance, the problems they address, and the caveats and pitfalls associated with their use.

5.1. Application Areas for A/B Decisions in Finance

(1) **Portfolio Optimization:** Bayesian A/B decision models are extensively used in portfolio optimization. By incorporating prior beliefs about asset returns and updating these beliefs with new data, investors can better estimate the expected returns and risks associated with different assets. This dynamic approach allows for continuous portfolio adjustment in response to new information, enhancing the ability to achieve optimal asset allocation.

(2) **Risk Management:** In risk management, Bayesian models provide a probabilistic framework for assessing the likelihood of various risk events. By combining historical data with expert judgment, these models can produce more accurate risk assessments. For instance, Bayesian models can be used to estimate the probability of default for a portfolio of loans, enabling more effective risk mitigation strategies.

(3) **Algorithmic Trading:** Algorithmic trading strategies often rely on Bayesian models to make real-time trading decisions. These models can incorporate prior knowledge and adapt to new market data, improving the accuracy of predictions regarding price movements. This adaptability is crucial in the fast-paced world of trading, where market conditions can change rapidly.

(4) **Credit Scoring:** Bayesian A/B decision models are also used in credit scoring to evaluate the creditworthiness of borrowers. By incorporating both historical data and expert judgment, these models can provide a more comprehensive assessment of a borrower's risk profile. This can lead to better lending decisions and reduced default rates.

(5) **Financial Forecasting:** Financial forecasting benefits from Bayesian models' ability to incorporate prior information and update forecasts as new data becomes available. This is particularly useful for predicting macroeconomic indicators, stock prices, and other financial metrics. The flexibility of Bayesian models allows for more accurate and reliable forecasts in the face of uncertainty.

5.2.   Advantages of Bayesian A/B Decision Models

Bayesian A/B decision models offer significant advantages over their frequentist counterparts in three areas: uncertainty, prior information, and dynamic updating.

(1) **Uncertainty in Data:** One of the primary problems addressed by Bayesian models is the inherent uncertainty in financial data. Traditional models often struggle with uncertainty, leading to less reliable results. Bayesian models, however, explicitly incorporate uncertainty into their framework, providing more robust and credible estimates.

(2) **Incorporating Prior Information:** In many financial contexts, prior information or expert knowledge is available but not utilized effectively in traditional models. Bayesian models allow for the incorporation of prior beliefs and expert judgment, which can enhance the model's performance, particularly when data is scarce or noisy.

(3) **Dynamic Updating:** Financial markets are dynamic, with conditions that can change rapidly. Bayesian models are particularly well-suited for dynamic environments because they continuously update their predictions as new data becomes available. This ensures that the models remain relevant and accurate over time.

5.3.   Caveats and Pitfalls in Bayesian Approaches

The analyst must keep aware of specific problems involved in Bayesian A/B tests, many of which are also shared in one form or another by frequentist tests.

(1) **Overfitting:** Like any model, Bayesian models are susceptible to overfitting, especially when too many parameters are estimated relative to the amount of data available. Overfitting can result in models that perform well on historical data but poorly on new, unseen data. Careful model validation and the use of techniques such as cross-validation can help mitigate this risk.

(2) **Interpretability:** Bayesian models can be complex and difficult to interpret, particularly for stakeholders who are not familiar

with Bayesian statistics. This can make it challenging to communicate the results and implications of the model to decision-makers. Efforts to simplify and explain the models, as well as the results, are necessary to ensure their effective use. In general, interpreting and communicating results of an analysis are simpler using Bayesian approaches.

(3) **Choice of Priors:** The selection of prior distributions is a critical aspect of Bayesian modeling. Poorly chosen priors can lead to biased results, undermining the model's credibility. It is essential to choose priors that are well-justified and reflect the true underlying distribution of the data.

(4) **Computational Complexity:** Bayesian models can be computationally intensive, particularly for large datasets or complex models. This can lead to long processing times and the need for significant computational resources. Advances in computational methods and software have mitigated this issue to some extent, but it remains a consideration.

(5) **Sensitivity to Assumptions:** Bayesian models rely on certain assumptions, such as the form of the prior distribution and the likelihood function. These assumptions can significantly impact the results, and incorrect assumptions can lead to misleading conclusions. It is crucial to test the sensitivity of the model to different assumptions and ensure that they are reasonable and well-supported by the data.

Bayesian A/B decision models offer a powerful and flexible framework for addressing various challenges in financial applications. Their ability to incorporate prior information, handle uncertainty, and dynamically update predictions makes them well-suited for the fast-paced and uncertain nature of financial markets. However, the effective application of these models requires careful consideration of several factors, including the choice of priors, computational complexity, the risk of overfitting, interpretability, and sensitivity to assumptions.

Financial professionals and analysts must be aware of these caveats and pitfalls to leverage the full potential of Bayesian A/B decision models while avoiding common mistakes. With proper implementation and ongoing evaluation, these models can significantly enhance decision-making processes, leading to better investment strategies, improved risk management, and more accurate financial forecasts. As the financial industry continues to evolve, the role of Bayesian models is likely to grow, providing valuable insights and driving more informed decisions in an increasingly complex and dynamic environment.

5.4. Case Study: CryptoBros', Inc. Portfolio Choice

The investment firm CryptoBros, Inc. was founded by two friends, Brendon and Trecool, to pursue arbitrage opportunities in video game gear and cryptocurrencies. Up to this time, they had considered A/B testing to be irrelevant to their specific business; they felt it was something that was used by marketers who needed to make a simple choice between different advertising campaigns or marketing strategies. In such situations, figures of merit involve customer engagement, click-throughs, or other non-financial measures, and marketers will not have accurate or complete information on the financial impact of their programs. Decisions in marketing are easily structured around a frequentist choice model for these reasons.

More recently, however, CryptoBros has been looking into the advantages offered through Bayesian approaches. Bayesian A/B testing opens up the possibility of inference on underlying profit drivers in a business, where risk, loss or profit functions can be calculated from the posterior. In financial portfolio choice, the data is typically in the form of a time series of prices of fungible assets, or functions of those prices, like return on investment (ROI).

CryptoBros were interested in one particular portfolio choice: whether to invest in the cryptocurrencies Bitcoin or Ether. Bitcoin, abbreviated BTC was the first decentralized cryptocurrency. Nodes in the peer-to-peer bitcoin network verify transactions through cryptography and record them in a public distributed ledger, called a blockchain, without central oversight. Consensus between nodes is achieved using a computationally intensive process based on proof of work, called mining, that guarantees the security of the Bitcoin blockchain. Ether, abbreviated ETH is the native cryptocurrency of the Ethereum platform. Ethereum is a decentralized blockchain with smart contract functionality. Among cryptocurrencies, Ether is second only to Bitcoin in market capitalization.

Both cryptocurrencies are widely traded, and have steadily increased in value over their lives. Bitcoin has a current market capitalization around $1.4 trillion and Ether around $460 billion. Both have made many people rich, and are sought after as investments. The question facing CryptoBros as erstwhile investors in cryptocurrencies was: "Which is the better investment?" In this case they interpreted "better" in terms of a ROI objective. Assuming they had some fixed budget which they will entirely invest in either Bitcoin or Ethereum. Which of these will make CryptoBros richer?

They began by extracting daily "high" prices for Bitcoin and Ether from Yahoo! Finance between November 9, 2017, and May 26, 2024, for 2391 observations of each price. From these, they computed the daily $ROI = \frac{price_t - price_{t-1}}{price_t}$ for each. Figure 5.1 graphs the density functions for these daily ROIs.

In this example, a choice of A means that the CryptoBros invest all their money in Bitcoin, and B in Ether. Here are the probabilities computed by `bayesTest` from our dataset, assuming an inverse-Normal-gamma prior with mean 0.0 ROI and standard deviation 0.05

(1) $P(A > B)$ by $(0, 0)$%: $\mu = 0.35952$ and $\sigma^2 = 0.0$
(2) The 5% and 95% Credible Interval on $(A - B)/B$ for interval length(s) $(0.9, 0.9)$: $\mu \in [-0.8719459, 1.8666860]$ with $\sigma^2 \in [-0.3510897, -0.2576500]$. Note that I have used the term *Credible Interval* rather than *Confidence Interval* (a concept from

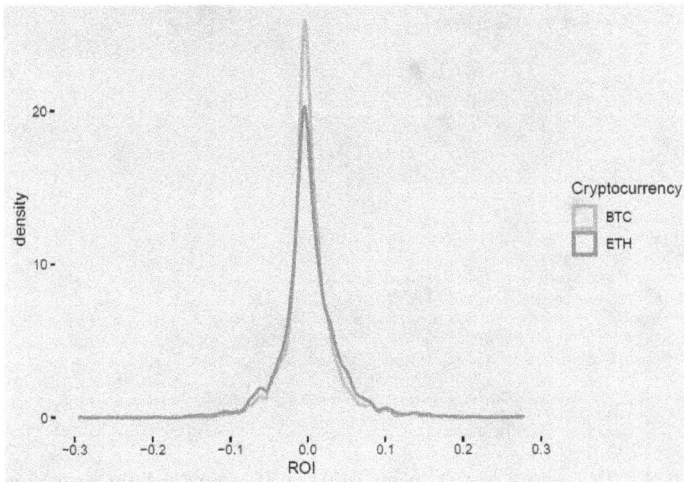

Figure 5.1. ROI density function for BTC (*A* light gray) and ETH (*B* dark gray).

fiducial inference) because posteriors graph *epistemic* probabilities, i.e., credences which are degrees of belief.

(3) The Posterior Expected Loss for choosing *A* over *B*: $\mu = 2.03662$ with $\sigma^2 = 0.4416585$.

Figures 5.2 and 5.3 graph the posterior means from `bayesTest`. Now consider each of these reported statistics in terms of their implications for the investor's portfolio choice.

- The first finding, that $P(A > B)$ only 36% of the time, implies that ROI will be better for Ether investments 64% of the time.
- The second finding reports the scaled difference between choices *A* and *B*, and the expectation of these will range between [−87%, 187%] implying a significant amount of volatility, which we would expect from the cryptocurrency markets. There is no underlying asset for cryptocurrencies and possibly a significant amount of rumor and market manipulation, all of which lend uncertainty to the market.
- The third finding suggests the investor will face an expected loss of 204% from choosing to invest in Bitcoin rather than Ether.

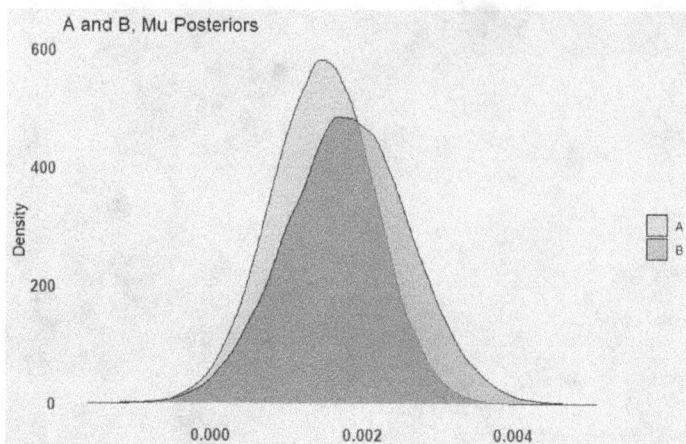

Figure 5.2. Posteriors of the mean ROI for Bitcoin (*A* light gray) and Ether (*B* dark gray).

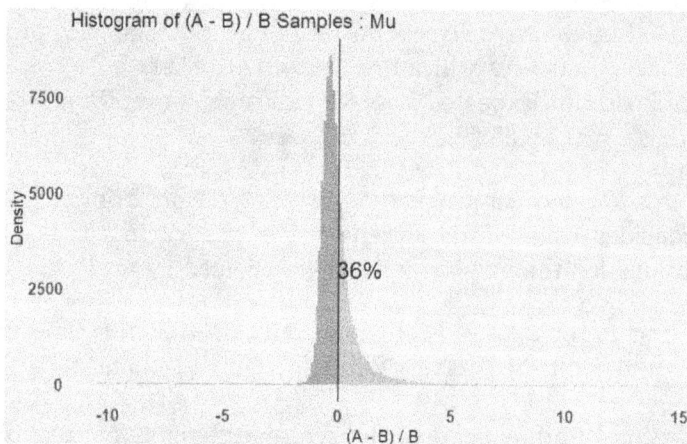

Figure 5.3. Posterior of the scaled difference of mean ROI for Bitcoin (*A* light gray) and Ether (*B* dark gray).

5.5. Case Study: Locust Lane Investments Explores National Economic Performance with a Decision Tree of A/B Tests

Locust Lane Investments is a global investment company that seeks long-run opportunities across a wide range of national economies. It recently reviewed a dataset from [32] of an annual return index for the invested global multi-asset market portfolio. The dataset covers the entire market of financial investors and analyzes returns and risk from 1960 to 2017. During that period, the market portfolio realized a compounded real return in US dollars of 4.45%, with a standard deviation of annual returns of 11.2% and a compounded excess return of 3.39%.

Locust Lane's dataset contained six asset price series that represented potential instruments for investing in the economy. Thus a binary A/B test was insufficient for a full analysis. They chose to perform a series of A/B tests (Figure 5.4) to see which sector of the economy has yielded the highest historical ROI using a series of pairwise comparison A/B tests in a decision tree (Figure 5.5), following the decision tree models in [33–35].

Locust Lane wanted to determine the best investment through a binary tree of A/B tests, each one selecting a "winner" from options A and B; that winner will then go on to be A/B tested against the next asset class as shown in the schematic below (Figure 5.5).

Densities of Asset Price ROIs

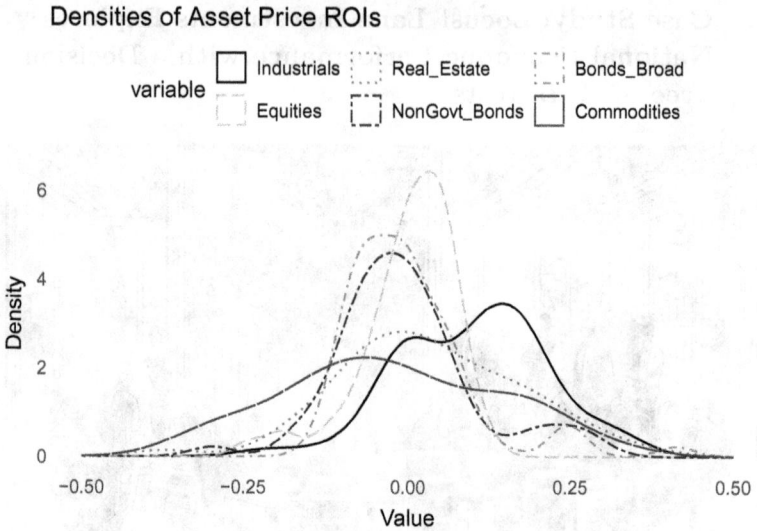

Figure 5.4. Density functions of asset class price ROI.

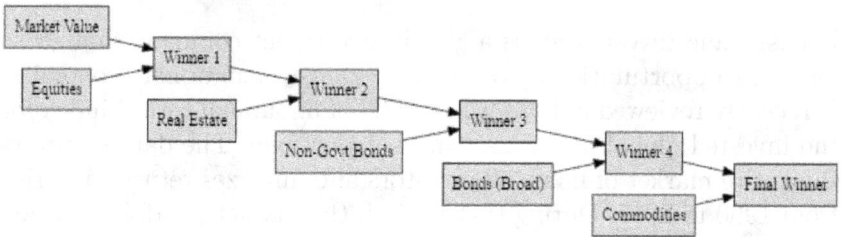

Figure 5.5. Locust Lane's decision tree of A/B tests.

Decision trees are an approach to decisions with more than two choices that have been successfully used in real options analysis for at least two decades. In Chapter 10, I present a more elegant, flexible and powerful approach to multiple choice models using multi-armed bandits. The results of their analysis are summarized in Tables 5.1 and 5.2.

The comprehensive analysis undertaken by Locust Lane Investments revealed several key insights that would shape its investment strategy for years to come. Through the meticulous examination of historical data and the implementation of sophisticated decision tree

Table 5.1. Fit statistics of each asset's time series.

Asset	Mean	SD	Skewness	Kurtosis
Commodities	0.032	0.281	2.281	10.288
Bonds_Broad	0.004	0.100	1.883	7.809
NonGovt_Bonds	0.011	0.147	2.395	13.438
Real_Estate	0.036	0.142	0.014	3.366
Equities	0.001	0.073	−1.278	5.100
Industrials	0.101	0.112	−0.184	3.172

Table 5.2. The asset class with the highest ROI.

Asset	ROI
Industrials	0.09948

models, the firm was able to identify sectors with the most promising returns and understand the underlying dynamics that drove these results.

First and foremost, the historical analysis underscored the importance of diversification. While individual sectors showed varying degrees of profitability, the data highlighted that a well-balanced portfolio across multiple sectors could effectively mitigate risks and enhance overall returns. This reinforced the company's commitment to maintaining a diversified investment strategy, ensuring that they would not be overly exposed to the volatility of any single sector.

The decision tree analysis revealed that certain sectors consistently outperformed others in terms of ROI. Technology, healthcare, and consumer goods emerged as the top performers, demonstrating robust growth and resilience over the decades. The technology sector, in particular, showed exceptional promise, driven by continuous innovation and increasing global demand for technological advancements. Healthcare also stood out due to its steady demand and the ever-growing need for medical services and products. Consumer goods, while slightly more volatile, showed strong returns driven by consistent consumer spending and brand loyalty.

Conversely, sectors such as utilities and energy, while offering stability, provided lower returns compared to their counterparts. The energy sector, in particular, was noted for its susceptibility to geopolitical tensions and fluctuating oil prices, which introduced significant uncertainty. Utilities, while stable, lacked the growth potential seen in more dynamic sectors.

Armed with these insights, Locust Lane Investments began to reallocate its resources strategically. The firm increased its investments in technology, healthcare, and consumer goods, capitalizing on their historical performance and future potential. At the same time, they reduced their exposure to the more volatile energy sector and the lower-growth utilities sector. This reallocation aimed to optimize the balance between risk and return, ensuring sustainable growth for the company's portfolio.

The analysis also highlighted the importance of geographic diversification. While the global market as a whole provided a solid return, certain regions outperformed others. North America and Asia, particularly China and India, demonstrated significant growth potential. These regions were characterized by rapid technological advancements, burgeoning middle classes, and favorable economic policies. On the other hand, some European markets showed slower growth, affected by political instability and economic challenges. Locust Lane Investments took these regional dynamics into account, increasing its investments in high-growth regions while carefully monitoring and adjusting its exposure to more volatile areas.

In addition to sector and geographic diversification, the analysis emphasized the value of adaptive investment strategies. Historical data showed that market conditions and sector performances evolved over time, influenced by technological advancements, regulatory changes, and macroeconomic trends. As a result, the firm recognized the need for flexibility in its investment approach, allowing for adjustments based on emerging trends and market signals. This adaptive strategy would enable Locust Lane Investments to stay ahead of the curve, capitalizing on new opportunities while mitigating potential risks.

The decision tree models also provided valuable insights into the risk-return profiles of different asset classes. Equities, while more volatile, offered higher returns compared to fixed-income securities. Real estate, on the other hand, provided a balanced mix of stability and growth, making it an attractive option for long-term investments. These findings led the firm to fine-tune its asset allocation, increasing its exposure to high-performing equities and strategically incorporating real estate into its portfolio.

Moreover, the analysis underscored the importance of continuous monitoring and evaluation. The financial landscape is dynamic, with new trends and challenges emerging regularly. Locust Lane Investments recognized that a static approach would not suffice in such an environment. Therefore, the firm committed to ongoing analysis and adjustment of its investment strategy, leveraging the latest data and analytical tools to make informed decisions. This proactive approach would ensure that the company remains well-positioned to navigate market fluctuations and capitalize on new opportunities.

The extensive analysis conducted by Locust Lane Investments provided a wealth of insights that significantly influenced its investment strategy. By identifying high-performing sectors, understanding regional dynamics, and emphasizing diversification and adaptability, the firm was able to optimize its portfolio for long-term growth and stability. The decision to increase investments in technology, healthcare, and consumer goods, while strategically diversifying across high-growth regions, positioned the company for sustained success in the global market.

Rigorous analysis of historical data and the implementation of advanced decision tree models provided Locust Lane Investments with a robust foundation for making informed investment decisions. The firm's strategic reallocation of resources, emphasis on diversification, and commitment to adaptability underscored its dedication to achieving sustainable growth. As Locust Lane Investments continued to evolve and refine its investment strategy, it remained confident in its ability to thrive in the ever-changing global market, securing a prosperous future for itself and its investors.

Chapter 6

Marketing Applications of A/B testing

A/B testing, a cornerstone of modern marketing, is a powerful method for comparing two versions of a marketing asset to determine which performs better. This method, also known as split testing, involves dividing an audience into two groups and exposing each group to a different version of a variable, such as an email subject line, website design, or advertisement. The goal is to identify which version drives higher engagement, conversions, or other desired outcomes. While A/B testing offers significant advantages in optimizing marketing strategies, it also comes with challenges and potential pitfalls. This narrative explores the application areas, problems addressed, and caveats and pitfalls associated with A/B testing in marketing.

6.1. Application Areas for A/B Decisions in Marketing

(1) **Website Optimization:** A/B testing is widely used in website optimization to enhance user experience and increase conversion rates. Marketers can test different versions of web pages, including headlines, call-to-action buttons, images, and layouts, to determine which variations lead to higher engagement and conversions. For example, an e-commerce site might test two different checkout processes to see which one reduces cart abandonment rates.

(2) **Email Marketing:** Email marketing campaigns benefit greatly from A/B testing. Marketers can test different subject lines, email copy, images, and call-to-action buttons to identify which elements improve open rates, click-through rates, and overall campaign effectiveness. By continuously optimizing email components, businesses can significantly enhance their email marketing performance.

(3) **Advertising Campaigns:** In advertising, A/B testing helps determine the most effective ad creatives, copy, and targeting strategies. Whether it's display ads, social media ads, or search engine marketing, A/B testing allows marketers to compare different ad variations to see which ones generate the highest click-through rates, engagement, and conversions. This data-driven approach helps optimize advertising spend and improve return on investment (ROI).

(4) **Content Marketing:** A/B testing is also crucial in content marketing. Marketers can test different types of content, headlines, and formats to understand what resonates best with their audience. For instance, a blog might test two different article titles to see which one attracts more readers. This helps in creating content that is more engaging and effective in achieving marketing goals.

(5) **Product Development:** A/B testing extends beyond traditional marketing tactics and into product development. Companies can test different product features, designs, and pricing models to determine which ones are more appealing to customers. This approach helps in making data-driven decisions that enhance product market fit and customer satisfaction.

6.2. Advantages of Bayesian A/B Decision Models

Bayesian A/B decision models offer significant advantages over their frequentist counterparts in three areas: optimization, consumer preferences and risk.

(1) **Optimization of Marketing Efforts:** A/B testing addresses the problem of optimizing marketing efforts by providing concrete data on what works and what doesn't. Instead of relying on

assumptions or intuition, marketers can make informed decisions based on empirical evidence, leading to more effective marketing strategies.

(2) **Understanding Customer Preferences:** A/B testing helps marketers understand customer preferences by revealing which variations of a marketing asset are more effective. This insight is crucial for creating personalized and relevant marketing campaigns that resonate with the target audience.

(3) **Reducing Risk:** By testing different variations on a smaller scale before a full rollout, A/B testing reduces the risk of failure. Marketers can identify potential issues and make necessary adjustments before committing significant resources to a particular strategy or campaign.

(4) **Improving Conversion Rates:** A/B testing directly contributes to improving conversion rates by identifying the most effective elements of a marketing campaign. Whether it's a more compelling call-to-action or a more engaging ad creative, A/B testing helps in fine-tuning marketing tactics to maximize conversions.

6.3. Caveats and Pitfalls in Bayesian Approaches

The analyst must keep aware of specific problems involved in Bayesian A/B tests, many of which are also shared in one form or another by frequentist tests.

(1) **Sample Size and Statistical Significance:** One of the most critical caveats of A/B testing is ensuring an adequate sample size and achieving statistical significance. Testing with too small a sample can lead to inconclusive or misleading results. Marketers must ensure that their tests run long enough and include a sufficiently large sample size to produce reliable results.

(2) **Testing Too Many Variables:** Testing too many variables at once can complicate the analysis and make it difficult to determine which changes influenced the outcome. It's essential to test one variable at a time or use multivariate testing if multiple variables need to be tested simultaneously.

(3) **Misinterpreting Results:** Misinterpreting the results of an A/B test is a common pitfall. It's crucial to understand that

correlation does not imply causation. Marketers must carefully analyze the data and consider other factors that might have influenced the results, such as seasonality or external events.

(4) **Focusing on Short-Term Gains:** A/B testing often focuses on short-term gains, such as immediate increases in click-through rates or conversions. However, it's important to consider the long-term impact of changes. For instance, a more aggressive call-to-action might boost short-term conversions but harm brand perception over time.

(5) **Ignoring the Bigger Picture:** A/B testing can lead to optimization in silos if not integrated with broader marketing strategies. It's essential to ensure that the insights gained from A/B testing align with overall business goals and marketing objectives. Focusing solely on optimizing individual elements without considering their impact on the overall customer journey can lead to suboptimal outcomes.

(6) **Implementation Challenges:** Implementing A/B tests can be challenging, especially for organizations with limited technical resources. Setting up tests, tracking results accurately, and making data-driven decisions require a certain level of expertise and technical capability. Without the right tools and skills, A/B testing efforts can be ineffective or counterproductive.

Bayesian A/B testing is a powerful tool in the marketer's arsenal, offering a systematic approach to optimizing various aspects of marketing strategies. By providing concrete data on what works and what doesn't, A/B testing helps marketers make informed decisions, understand customer preferences, and reduce the risk of failed campaigns. Its applications span across website optimization, email marketing, advertising, content marketing, and even product development, making it a versatile and valuable methodology.

However, to harness the full potential of A/B testing, marketers must be aware of its caveats and pitfalls. Ensuring adequate sample sizes, avoiding the testing of too many variables at once, interpreting results correctly, and considering both short-term and long-term impacts are critical for successful A/B testing. Additionally, integrating A/B testing insights with broader marketing strategies and having the necessary technical capabilities are essential for effective implementation.

By navigating these challenges thoughtfully and strategically, marketers can leverage A/B testing to enhance their marketing efforts, drive higher engagement and conversions, and ultimately achieve their business objectives. In an increasingly data-driven marketing landscape, mastering the art and science of A/B testing is more important than ever for staying competitive and delivering exceptional value to customers.

6.4. Case Study: Impact of User Interface Design on LuxeHome Decor's Product Sales

This case study examines the influence of user interface (UI) modifications on consumer purchasing behavior within a fictional mid-sized company, LuxeHome Decor. LuxeHome Decor, known for its eclectic mix of home furnishings, has recently been interested in maximizing their online sales through strategic enhancements to their e-commerce platform. This exploration led them to conduct an A/B testing to determine if a redesigned UI could significantly improve product sales.

LuxeHome Decor has been experiencing steady growth but noticed that the conversion rate on their e-commerce site has plateaued despite increased traffic. The marketing team hypothesized that the current UI might be affecting customer purchase decisions. To verify this, they decided to initiate an A/B test comparing the current UI with a new, more streamlined version designed to enhance user experience and simplify the purchasing process. The primary objective of the study was to determine whether the alternative user interface could increase sales compared to the current design.

LuxeHome Decor's e-commerce team decided to run a two-week A/B test on their most popular product, the "Luxe Lamp". They randomly assigned visitors to their website into two groups:

- **Group A (Control):** Users in this group interacted with the product through the default user interface.
- **Group B (Treatment):** Users in this group experienced the redesigned interface. Both groups consisted of 100 users to maintain statistical relevance, and the outcome of interest was the number of purchases made from each group.

The results after two weeks were as follows:

- **Group A:** 45 purchases out of 100 interactions.
- **Group B:** 49 purchases out of 100 interactions.

Analysis of the data showed that it followed a binomial distribution, with each user decision modeled as a Bernoulli trial where a purchase was marked by 1 and no purchase by 0.

To determine if the observed difference in conversion rates between the two interfaces was statistically significant, the marketing department performed a hypothesis test to implement a traditional frequentist A/B model. The null hypothesis (H_0) stated that there was no difference in purchase rates between the two interfaces, while the alternative hypothesis (H_A) suggested that the new UI led to a higher purchase rate.

The p-value calculated from the test was 0.453, indicating that the differences observed were not statistically significant at the conventional 0.05 significance threshold. This meant that while there was an improvement in the purchase rate with the new UI, it was not enough to conclusively say it was due to the interface redesign.

After frequentist A/B tests failed to provide the company with any strategic guidance, it decided to experiment with a Bayesian A/B decision model. Data had already been collected, and decision-making was merely a matter of applying the data they had already collected.

The marketing team started by conducting a Bayesian test using just website impressions and purchases. This replicated the setup they had used for the frequentist test, while allowing for a more flexible set of distributions to fit their dataset. Once again, they wanted to know whether a product sells better with the new change to the online user interface, and they ran an experiment on the two groups:

- **Group *A* (control):** 45 users out of a sample of 100 purchased the product with the default user interface.
- **Group *B* (treatment):** 49 users out of a sample of 100 purchased the product with the alternative user interface.

They assumed that the data was binomially distributed, giving a Bernoulli sequence of impressions versus purchases. An impression without a purchase is a 0 and with a purchase is a 1.

In contrast to their frequentist model, this time, with Bayesian A/B testing, their results determined that the treatment user interface (Group B) would lead to a larger number of purchases around 76% of the time.

This result (Figure 6.1) not only provided clear direction on whether their attempts to improve the user interface were effective,

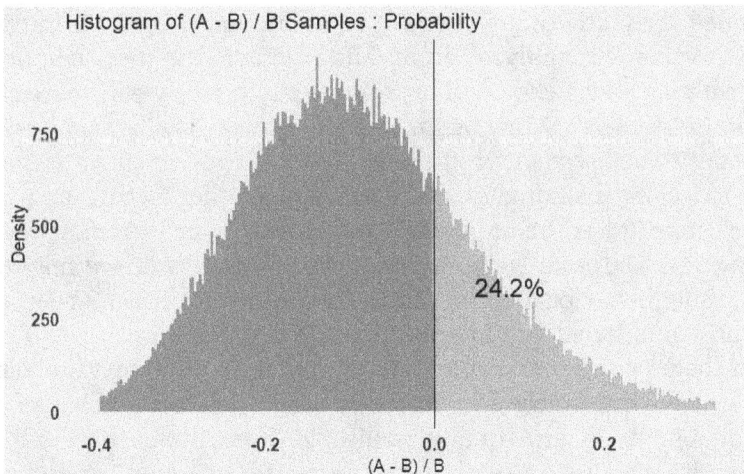

Figure 6.1. The new UI (group *B* in dark gray) will increase the sales of the Luxe Lamp 76% of the time.

but they provided a significant satisfaction to the group members assigned to present their results: interpretability. Interpretability seems like an abstract concept, but to LuxeHome Decor the satisfaction was tangible. Previously during the meeting in which they presented their results, they had watched eyes glaze over, and fielded derision and "Why are we paying you?" questions when they explained

p-value calculated from the test was 0.453 and we couldn't tell you whether the new UI was effective.

as they struggled to convince the company that they knew what they were doing. But the results of Bayesian A/B tests fit into an easy narrative about more purchases, i.e.,

our results determined that the treatment user interface (group B) would lead to a larger number of purchases around 76% of the time.

Furthermore, they promised that with financial data that they would be able to specify precisely how profitable business would be under the two competing UIs. Additionally, they could also generate risk profiles that would be useful for maintaining proper inventory levels to support customer purchases, and assure that shipments could be made promptly.

LuxeHome Decor's management was impressed enough that they provided the marketing team with an additional budget for Bayesian A/B testing to analyze profitability under the two competing Determining which choice, A or B, of user interface yields more clicks per impression is useful, but ultimately we would like our business decisions to be based on profitability. In most experiments it is impossible to isolate the influence of treatments on other factors, and these factors may indeed be important to us, even though we may not initially consider them. Here, a change in the user interface may alter purchasing behavior more generally, and cause buyers in group A to purchase a different product mix from group B.

In the previous test of alternative user interfaces they were merely concerned with which user interface lead to more purchases. But not all purchases are equally profitable. Some items may be loss-leaders to encourage use of an online platform; some items may be

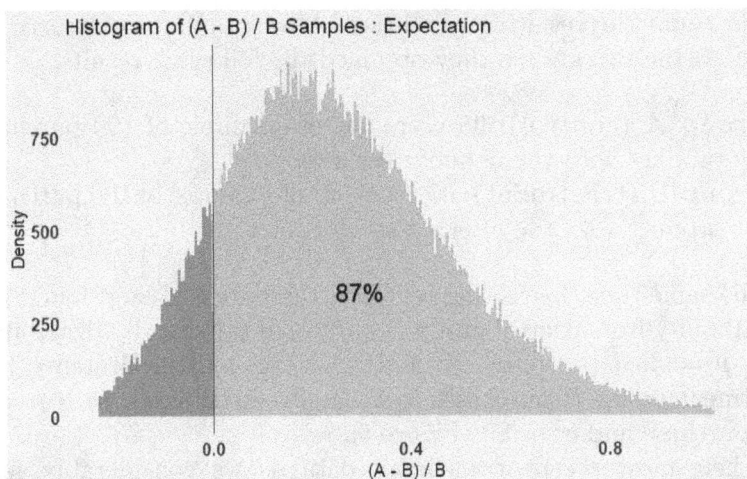

Figure 6.2. After considering profitability of purchases, the marketing team reversed their previous decision, and decided that the original user interface was preferable.

discounted; and so may be faddish and generate significant profits. Assume that marketing has found that profitability from control group A purchases is Normally distributed, with $\mu = \$6$ per purchase and $\sigma^2 = \$1$, and profitability from treatment group B purchases is Normally distributed, with $\mu = \$5$ per purchase and $\sigma^2 = \$5$. Then they can use the `combine` function in the `bayesAB` package to determine the user interface with higher expected profitability.

Once they considered the profitability of purchases, their decision changed, and they concluded that the original user interface (group A) would lead to a more profit around 87% of the time (Figure 6.2).

6.4.1 *Updating with a new test*

LuxeHome Decor was so pleased with the results and interpretability of these Bayesian A/B decision models that the planned to repeat the experiment on a regular basis, and pool the results of all of the tests. The would collect observations, stopping after both the control A and the treatment B user interface received 100 impressions, then they would count the resulting purchases. After running the experiment again on the two groups they obtained the following results:

- **Group A (control):** 35 users out of a sample of 100 purchased the product with the default user interface.
- **Group B (treatment):** 30 users out of a sample of 100 purchased the product with the alternative user interface.

At the same time, marketing reviewed the purchases and found that profitability from control group A purchases is Normally distributed, with $\mu = \$5.5$ per purchase and $\sigma^2 = \$4$, and profitability from treatment group B purchases is Normally distributed, with $\mu = \$5$ per purchase and $\sigma^2 = \$3$ (Figure 6.3).

Their most recent tranche of data shows considerably fewer purchases per 100 impressions than the first tranche of observations, while the profitability of the original group A purchases is slightly higher (Figure 6.3). After updating the previous posterior with the new observations, they concluded that the treatment user interface (group B) would lead to a more profit around 63% of the time.

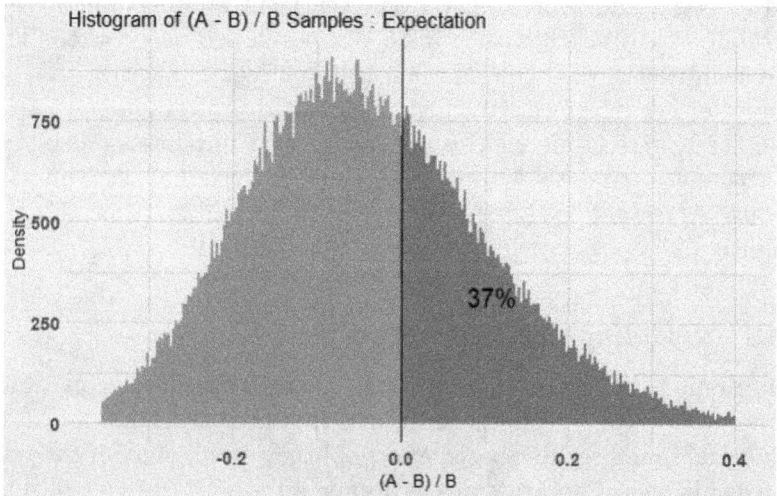

Figure 6.3. After collecting additional data which also considered profitability of purchases, the marketing team once again reversed their previous decision, and decided that the new user interface was preferable.

6.5. Case Study: Google Analytics at Dashuju.com

Companies today have access to a wealth of data that they could only dream of even a decade ago. But with that information bounty comes a new challenge: using that data for competitive advantage. mastering the art of analytics is crucial for any company aiming to stay competitive. *Dashuju.com* was one such company; a major e-commerce platform that sought to leverage data analytics to refine its marketing strategies and drive sales growth. Dashuju's journey unfolds as the analytics team employs advanced statistical analysis to dissect user behavior and purchasing patterns, providing actionable insights that would shape the company's future marketing initiatives.

Dashuju, facing inconsistent sales and varying user engagement metrics, sought to understand the underlying factors better. The goal was clear: to optimize marketing tactics in order to enhance conversion rates and raise average purchase values through a deep dive into the data collected via Google Analytics.

They began with the meticulous compilation of user interaction data from Google Analytics and purchase histories stored in an extensive corporate database. The analytics team began by processing and categorizing this data to identify distinct patterns and trends.

A pivotal step in the analysis was to segment the users into two main groups: *new visitors* and *returning visitors*. This categorization was based on the premise that these groups exhibited different behaviors and preferences which, if understood, could be capitalized on to tailor marketing strategies effectively.

The team first analyzed the session behaviors of returning visitors. By examining the frequency and recency of their visits, insights into their engagement levels were gleaned. These metrics provided clues about the effectiveness of current retention strategies and highlighted potential areas for improvement.

The marketing department at Dashuju.com felt that returning visitors were the segment that mainly drove sales, reasoning that because they returned again and again, they were satisfied with the products and continued to be loyal customers. Marketing tended to dismiss new visitors as "window shoppers" who consumed bandwidth on the site, but didn't deliver the sales.

Dashuju.com's platform managers were agnostic. They certainly listened to their marketing department, but ultimately, they insisted on letting "the data speak for itself." To that end, they chose to employ Bayesian A/B tests. They collected Google Analytics site traffic information on visitors and sales between February 1, 2024, and May 31, 2024, retrieving 10,000 transactions.

Test 1 They split the dataset by the `userType` field into "new" and "returning" vistors. Their metric for comparisons was sales per visit.

Test 2 Dashuju.com's platform managers performed a second comparison for customers that looked at recency. They split the dataset by the `daysSinceLastSession` field into customers who returned within a two days, and those who returned in more than two days, with a comparison metric of sales per visit.

To compare the behaviors between new and returning visitors more rigorously, the team employed Bayesian A/B decision models that allowed them to understand with greater clarity how different segments reacted to various marketing tactics, thus providing a probabilistic assessment of potential strategies.

A preliminary analysis of new versus returning customers provided strong support for their intuition that sales were mainly driven by new customers (Figure 6.4).

Dashuju.com's management, though, was skeptical, feeling that the resolution offered by a simple new versus returning split was too crude to capture consumer behavior. They suggested that instead, customers who returned quickly after visiting Dashuju's platform

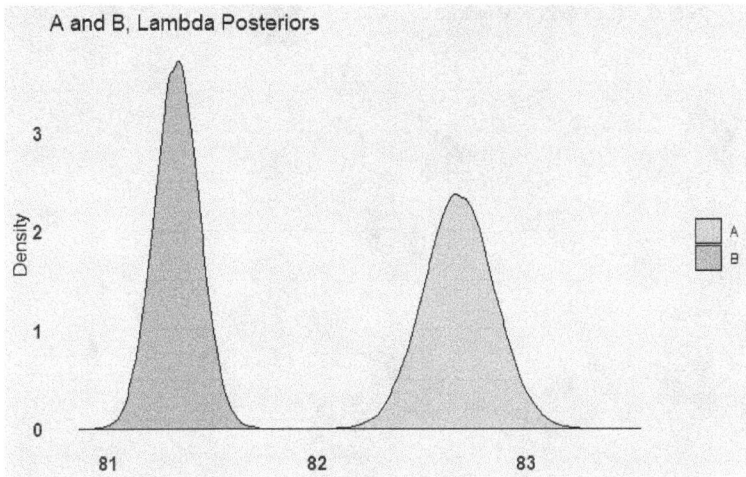

Figure 6.4. New (light gray) versus Returning (dark gray) visitors in generating Sales.

were probably interested and motivated to buy, but perhaps more thoughtful in their decisions. The suggested a second A/B test that would pit quickly returning visitors against those who returned at a more leisurely pace. They set the cutoff at two days, figuring that was a more appropriate way to interpret consumer behavior. Indeed the A/B testing showed that quickly returning customers spent about $1 less per sale than those who returned at a more leisurely pace (Figures 6.5 and 6.6).

Marketing was in turn skeptical, and suggested a synthesis – combining the two results using the `combine` function in the `bayesAB` R-language package. A decision-making consensus is easier when there is one performance statistic to present to management, compared to a suite of metrics. Thus, Marketing suggested that these individual A/B tests results would be more compelling were they combined. The suggested combining new customers with customers who returned within two days. Everything else would be grouped into the customers who eventually returned.

The results supported their initial insight that new and quickly returning customers were driving 95% of sales.

A critical part of the analysis was understanding purchasing patterns. The team analyzed the probability of purchases among the visitors and the typical spending amounts, employing models suited for

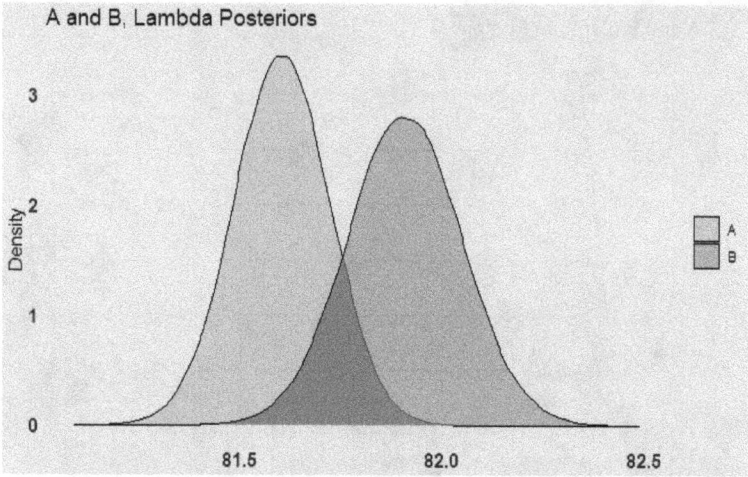

Figure 6.5. Quickly returning (light gray) versus Leisurely returns (dark gray) in generating sales.

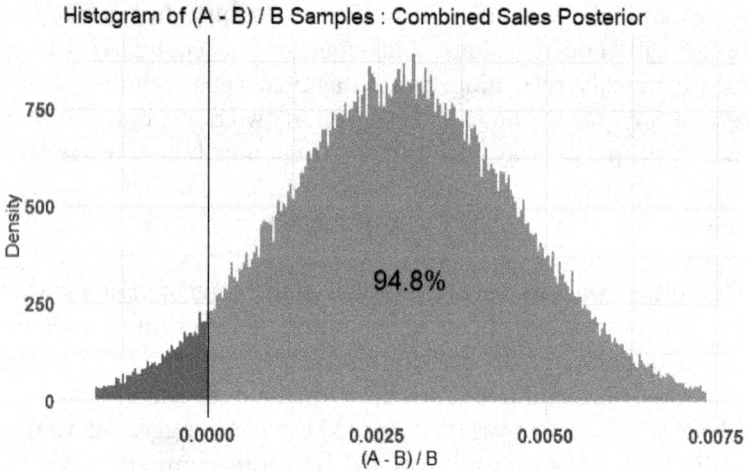

Figure 6.6. Marketing's Synthesis: Normalized Advantage of New (light gray) over Returning (dark gray) Customers in generating sales.

data with many non-purchasers. These patterns were crucial in identifying how much each segment was likely to spend, which directly influenced marketing and pricing strategies.

The final analytical step involved synthesizing session data with purchasing information to estimate potential sales revenues.

This comprehensive approach helped in constructing a more complete picture of how engagement levels correlated with purchasing behaviors, thereby enabling more accurate revenue forecasting.

The combined posterior of these two tests suggests that new and quickly returning customers generate ~95% more sales than customers who take longer to return. Marketing's considerable emphasis on "recency" of customer visits was supported by the data, both in the individual tests of new versus returning customers, as well as the quickly returning customers versus those who take longer to return.

From the insights gathered, the team identified clear differences between new and returning visitors — not just in their session behaviors but also in their purchasing tendencies. Visitors who took a long time to return were generally hesitant and spent less, while new visitors showed higher engagement and greater spending. This is what Marketing had originally suggested, and the data supported their intuition.

Based on these insights, the marketing department developed tailored strategies. For returning visitors, they implemented targeted email campaigns that leveraged existing engagement. For new visitors, efforts were directed toward enhancing the onboarding process to improve initial experiences and encourage first-time purchases.

Through meticulous data analysis, the company not only gained a deeper understanding of its customer base but also enhanced its strategic marketing initiatives and increased sales. This story exemplifies how leveraging data analytics can lead to targeted actions that significantly impact a company's bottom line.

Chapter 7

A/B Testing in Insurance, Warranties and Risk Management

Traditional frequentist A/B testing is not particularly useful for risk management applications, insurance or warranty and repair management because these applications deal with extreme values and tails of probability distributions rather than means. Thus A/B testing has not typically found an application in these areas. But Bayesian approaches, because they offer a full posterior distribution, not just a point estimate or a p-value, can, with proper interpretation, yield decision models for extreme situations.

7.1. Application Areas for A/B Decisions in Insurance, Warranties and Risk Management

(1) **Insurance Policy Design:** Insurance companies can use A/B testing to optimize policy design and pricing. By testing different versions of policy terms, coverage options, and pricing structures, insurers can identify the combinations that attract more customers and generate higher revenue. For example, an insurer might test two versions of a health insurance policy, one with lower premiums and higher deductibles and another with higher premiums and lower deductibles, to see which one customers prefer.

(2) **Marketing and Customer Acquisition:** A/B testing is widely used in marketing strategies for insurance products. Insurers can test different marketing messages, channels, and

campaigns to determine which ones are most effective in acquiring new customers. For instance, a company might test different email subject lines, landing page designs, or advertising copy to see which version leads to higher conversion rates.

(3) **Customer Retention and Engagement:** Insurance companies can use A/B testing to enhance customer retention and engagement strategies. By testing different approaches to communication, customer service, and loyalty programs, insurers can find the most effective ways to keep customers satisfied and reduce churn. For example, an insurer might test two different customer service scripts to see which one results in higher customer satisfaction scores.

(4) **Claims Processing:** A/B testing can be applied to optimize the claims processing workflow. By testing different methods of handling claims, insurers can identify the most efficient and customer-friendly approaches. For instance, an insurer might test a traditional claims process against a streamlined, digital-first approach to see which one leads to faster resolution times and higher customer satisfaction.

(5) **Risk Assessment and Underwriting:** In risk assessment and underwriting, A/B testing can help insurers refine their models and criteria. By testing different risk factors and scoring models, insurers can improve the accuracy of their risk assessments and underwriting decisions. For example, an insurer might test the impact of including additional data points, such as social media activity or health tracking data, in their risk assessment models.

(6) **Warranty Services:** For companies offering warranties, A/B testing can optimize the terms and conditions of warranties, as well as the process for filing and resolving claims. By testing different warranty terms, such as coverage duration and exclusions, companies can determine the most attractive and cost-effective options for their customers.

7.2. Advantages Offered by Bayesian A/B Decision Models

Bayesian A/B decision models offer significant advantages over their frequentist counterparts in three areas: optimization, user experience, operating costs and risk.

(1) **Optimization of Products and Services:** A/B testing addresses the challenge of optimizing insurance policies, warranties, and related services. By providing empirical data on customer preferences and behavior, A/B testing helps companies design products and services that better meet customer needs and expectations.

(2) **Enhancing Customer Experience:** A/B testing helps insurers and warranty providers enhance the customer experience by identifying the most effective ways to engage and retain customers. This includes optimizing communication strategies, customer service interactions, and claims processes.

(3) **Reducing Operational Costs:** By identifying more efficient processes and effective marketing strategies, A/B testing can help insurers and warranty providers reduce operational costs. For instance, testing different claims processing methods can reveal ways to streamline workflows and reduce administrative expenses.

(4) **Improving Risk Management:** A/B testing improves risk management by refining risk assessment and underwriting models. By testing different variables and criteria, insurers can enhance the accuracy of their risk predictions and make more informed underwriting decisions.

7.3.   Caveats and Pitfalls in Bayesian Approaches

The analyst must keep aware of specific problems involved in Bayesian A/B tests, many of which are also shared in one form or another by frequentist tests.

(1) **Sample Size and Statistical Significance:** One of the most critical caveats of A/B testing is ensuring an adequate sample size and achieving statistical significance. Small sample sizes can lead to inconclusive or misleading results. Insurers must ensure that their tests run long enough and include a sufficiently large sample size to produce reliable results.

(2) **Ethical and Regulatory Considerations:** In the insurance and warranty industries, ethical and regulatory considerations are paramount. A/B testing must be conducted in a way that respects customer privacy and complies with regulatory

requirements. For example, testing different underwriting criteria must not result in discriminatory practices or violate data protection laws.

(3) **Overlooking Long-Term Impacts:** A/B testing often focuses on short-term results, such as immediate increases in customer acquisition or satisfaction. However, it's important to consider the long-term impact of changes. For instance, a more aggressive marketing message might boost short-term sales but could harm the company's reputation over time if it leads to unrealistic customer expectations.

(4) **Implementation Challenges:** Implementing A/B tests in the insurance and warranty sectors can be challenging due to the complexity of products and services. Setting up tests, tracking results accurately, and making data-driven decisions require a certain level of expertise and technical capability. Without the right tools and skills, A/B testing efforts can be ineffective or counterproductive.

(5) **Misinterpreting Results:** Misinterpreting the results of an A/B test is a common pitfall. It's crucial to understand that correlation does not imply causation. Insurers must carefully analyze the data and consider other factors that might have influenced the results, such as seasonality or external events.

(6) **Testing Too Many Variables:** Testing too many variables at once can complicate the analysis and make it difficult to determine which change influenced the outcome. It's essential to test one variable at a time or use multivariate testing if multiple variables need to be tested simultaneously.

(7) **Risk of Bias:** Bias in A/B testing can lead to inaccurate results. Selection bias, response bias, and other forms of bias can skew the findings and lead to incorrect conclusions. Ensuring a randomized and representative sample is crucial for obtaining valid results.

Bayesian A/B testing is a powerful tool for optimizing insurance policies, warranties, and risk management strategies. By providing concrete data on what works and what doesn't, A/B testing helps insurers and warranty providers make informed decisions, enhance customer experience, and improve operational efficiency. Its applications span across policy design, marketing, customer retention, claims processing, risk assessment, and warranty services, making it a versatile and valuable methodology.

However, to harness the full potential of A/B testing, companies must be aware of its caveats and pitfalls. Ensuring adequate sample sizes, addressing ethical and regulatory considerations, avoiding the testing of too many variables at once, interpreting results correctly, and considering both short-term and long-term impacts are critical for successful A/B testing. Additionally, having the necessary technical capabilities and mitigating the risk of bias are essential for effective implementation.

By navigating these challenges thoughtfully and strategically, insurers and warranty providers can leverage A/B testing to enhance their products and services, drive higher customer satisfaction and retention, and ultimately achieve their business objectives. In an increasingly data-driven industry, mastering the art and science of A/B testing is more important than ever for staying competitive and delivering exceptional value to customers.

7.4. Case Study: The Cost of Pet Insurance

In a quaint neighborhood in a neat suburban village, two women, Sumar and Martha, shared more than just a love for golden retrievers.

They were both devoted pet owners, each with their unique approach to caring for their furry friends. Sumar's golden retriever, Dusty, was a bundle of joy with a shimmering coat that reflected her love and meticulous care. Martha's golden retriever, Noodles, was equally cherished but led a more laid-back lifestyle.

Sumar adored Dusty beyond measure. She pampered him with homemade meals crafted from the finest meats and freshest vegetables she could find at the farmer's market. To ensure Dusty's future was secure, Sumar invested in a $10,000 pet life insurance policy from Metropolitan Pets Mutual. This policy was meant to cover Dusty's veterinary expenses and cremation costs when his time eventually came. Despite knowing that golden retrievers typically lived only seven to eight years, Sumar held a steadfast belief that Dusty would defy the odds and live to be 20 years old.

Across the street, Martha cared deeply for Noodles but had a different philosophy. She believed that diet had little impact on a golden retriever's lifespan and fed Noodles dried dog food from the local store. Just like Sumar, Martha also bought a $10,000 pet insurance policy from Metropolitan Pets Mutual. And, despite her more relaxed approach to Noodles' diet, she too believed that her beloved pet would live to be 20 years old.

One sunny afternoon, the two women met over tea and discussed their concerns about their pets' health and longevity. Sumar, ever meticulous, was anxious about Dusty's health and wanted to ensure she was doing everything possible. Martha, calm and practical, shared her thoughts on how much a dog's diet truly mattered.

As their conversation unfolded, they decided to seek advice from Metropolitan Pets Mutual, the insurance company they both trusted. Metropolitan Pets Mutual had a wealth of data on golden retrievers and their lifespans, including insights into how diet might affect their longevity. They provided Sumar and Martha with detailed data, which helped them update their beliefs about Dusty's and Noodles' lifespans.

Using this data, Sumar and Martha performed an analysis to compare the net present value (NPV) of the health care and ultimately cremation claims that would be borne by their insurance policies, taking into account their different caregiving philosophies. They learned that Metropolitan Pets Mutual modeled the survival function for age using a lognormal distribution, a method relatively insensitive to prior beliefs.

As they pored over graphs and tables, they saw that despite their hope and care, the harsh reality was that golden retrievers generally lived only around seven years. Even though both women believed their dogs would live to 20, the data from Metropolitan Pets Mutual painted a different picture.

The analysis revealed that the meticulous care Sumar lavished on Dusty significantly extended his lifespan compared to Martha's straightforward approach with Noodles. Yet, both women found solace in knowing they were prepared and that their beloved pets would be taken care of no matter what.

As the sun set over their picturesque neighborhood, Sumar and Martha gathered once again on Sumar's porch, their golden retrievers, Dusty and Noodles, playing in the yard. They were eager to delve deeper into the data provided by Metropolitan Pets Mutual, hoping to uncover the truth behind their beloved pets' potential lifespans.

Sumar had spread out graphs and tables across the table, each one a piece of the puzzle they were trying to solve. She pointed to Figures 7.1 and 7.2 complied from their analysis of the insurance company's extensive statistics.

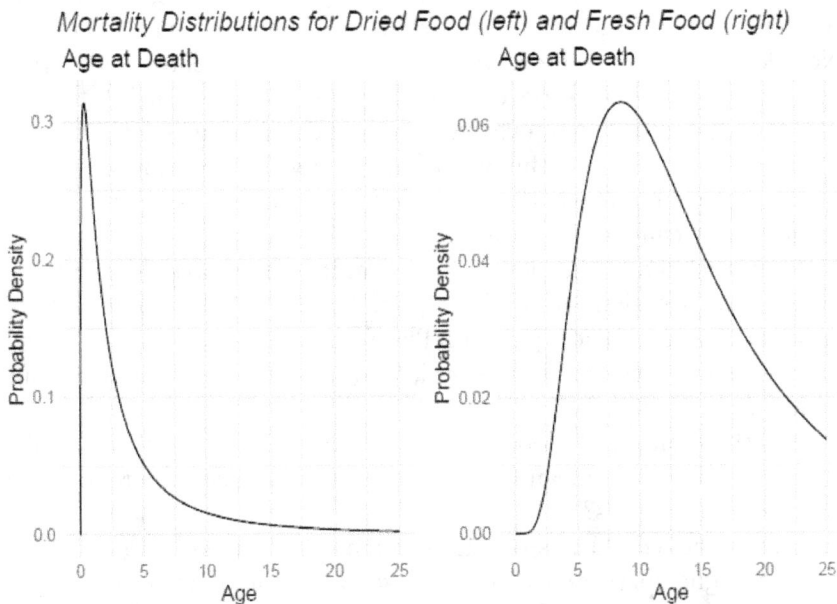

Figure 7.1. Mortality distributions for dried food (left) and fresh food (right).

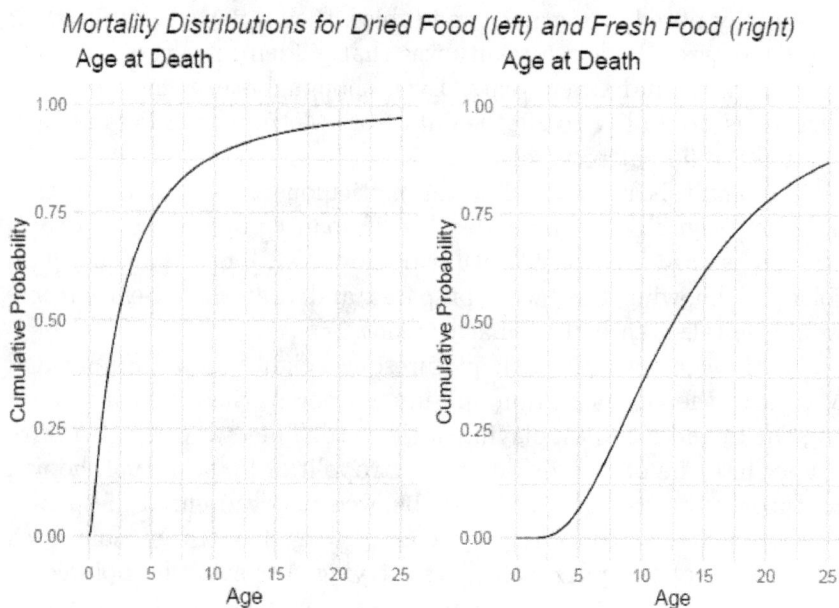

Figure 7.2.   Mortality distributions for dried food (left) and fresh food (right).

"Look at this, Martha," Sumar said, tracing her finger along one of the lines on the graph. "We started with the belief that Dusty and Noodles would live to be 20 years old. But the data shows a different reality. Most golden retrievers don't even make it to ten."

Martha nodded, her eyes scanning the information. "It's a bit disheartening, isn't it? But what caught my eye was how our prior beliefs hardly impacted the final analysis. It seems our initial optimism didn't change the outcomes much."

Sumar leaned back, thoughtful. "Yes, it underscores the prudence of Bayesian revision. By updating our beliefs based on actual data, we get a more realistic picture. The frequentist methods predicted more expensive claims, but Bayesian revision helped us manage our expectations better."

Martha glanced at Noodles, who was happily chasing a butterfly. "So, what does this mean for us? Do you think these differences will affect our premiums?"

Sumar sighed. "I suspect they might. If frequentist predictions lead to higher expected claims, those costs could be passed on to us as higher premiums. It's a bit of a double-edged sword. We want

Table 7.1. Comparative actuarial costs of pet insurance.

Diet	Frequentist NP cost of insurance claim ($)	Bayesian posterior NP cost of insurance claim ($)
Dried food	8245.22	7499.04
Fresh food	5186.66	4801.26

to be prepared for the worst, but we also don't want to overpay." (Table 7.1 and Figures 7.1 and 7.2).

The two women sat in silence for a moment, the sounds of their dogs playing filling the air. Despite the sobering analysis, they found comfort in knowing they were taking the right steps to protect their pets.

As they watched Dusty and Noodles, Sumar smiled. "No matter what the statistics say, we know we're doing our best for them. Whether it's through careful feeding or just lots of love and attention, we're giving them the best lives we can."

Martha returned the smile. "Absolutely. And if that means adjusting our expectations and being prepared for any outcome, then so be it. They're worth it."

In that shared understanding, Sumar and Martha found peace. Their journey through the maze of data and statistics had brought them closer, and their bond with Dusty and Noodles stronger. They knew that, regardless of the numbers, every day with their golden retrievers was a gift to be cherished.

7.5. Case Study: The Sexy Son

In the pulsating heart of the city's nightlife, where neon lights flicker and music throbs through the crowded bars and dance floors, a modern adaptation of an age-old evolutionary drama unfolds. Tonight, young Ron is again at his favorite watering hole among the 20-somethings navigating the dating scene. Casually, Ron swipes right on Tinder competitor *MatchRate*, the popular dating app where the *sexy son hypothesis* plays out every night in 21st-century courtship rituals.

The sexy son hypothesis was proposed by academics in the 1970s based on statistician Ronald Fisher's observations of mating preferences in birds. The sexy son hypothesis suggests a subtly sophisticated strategy that governs the mate choices among these young revelers. Young ladies impelled by atavistic urges scan the room for a partner whose genes promise strong, manly, charismatic sons — the kind who will thrive in the future dating markets. Or at least this is what our protagonist Ron has been told.

In this social jungle, the attributes that once might have included territory or nuptial gifts in the wild translate into charisma, style, or an infectious smile — traits that might predict a son's success in his own romantic endeavors. As Fisher's principle hints, the success of one's genes is measured not just by the quantity of offspring, but by their quality and desirability in such social contexts.

Females in these settings are not just passive participants; they are discerning selectors, influenced by a cascade of atavistic cues. They might ponder, perhaps not in so many words: "Who among these has the genetic flair to father sons who will charm, who will dance, who will swipe right into the hearts of their future matches?"

Thus, each choice at the bar, each swipe-right on MatchRate, is part of a complex dance driven by millennia of evolutionary strategy, distilled now into moments of flirtation and attraction. The dazzling array of male fashion, witty banter, and gym-toned physiques are modern feathers and dances, displays evolved not just for survival in the traditional sense, but for thriving in the social and romantic arenas of human life. But in the 21st century, all of these male cues have been algorithmically distilled into the *dating profile*.

And in this sultry jungle, surveying the bar, our protagonist Ron is deciding how to optimally craft his profile on MatchRate.

MatchRate quantifies life, optimally describes it in pithy, succinct profiles, and publishes these to potentially eligible partners, promising love with every interaction in every choice made. Beneath MatchRate's surface of social fun, the age-old narrative of natural selection scripts the scene, guiding the algorithms that will shape the generations to come. Here, in the vibrant whirl of the city's nightlife, the dance of evolution continues, as subtle and enduring as ever, turbocharged by artificial intelligence. MatchRate caters to those who seek the perfect "10", who love to rate their dates and enjoy receiving ratings in return, all in the quest for love.

Ron is torn: should he aspire to be the future father to Fisher's "Sexy Sons" or should he pose as "sensitive Ron," the doting future partner? He must decide between one of two profiles: *A* or *B*. Let's take a look.

Profile A: *Username:* IronRon

Tagline: Adventurer, charmer, and your future gym buddy.

About Me: I'm Ron, and I'm here to steal the spotlight in your daily routine. Standing tall, with a physique reminiscent of the legendary Arnold Schwarzenegger, I bring a blend of brawn and beauty that's sure to catch your eye. My friends describe me as "handsomely rugged", and I pride myself on maintaining an athletic lifestyle that keeps me in top form — both physically and mentally.

What I'm looking for: I'm seeking someone who appreciates witty banter and spontaneous adventures. If you enjoy intellectual conversations that are as engaging as a night out or a cozy evening by the fireplace, we might just be the perfect match. I'm attracted to women who are confident, kind, and have a zest for life.

Hobbies:

- Hitting the gym (early bird catches the worm!).
- Hiking in the great outdoors (nature's gym!).
- Cooking protein-packed meals (fuel for the body and soul).
- Reading everything from Nietzsche to Neil Gaiman.

Fun Facts:

- I can quote every line from "The Godfather."
- I've been known to serenade the occasional karaoke crowd with a surprisingly accurate Elvis Presley impression.
- I brew my own beer, which pairs perfectly with my legendary BBQs.

Why You Should Swipe Right: With me, you'll find a partner who is not only confident in his skin but also genuinely passionate about forming a real connection. My sense of humor is as strong as my bench press, and if you appreciate a man who can both challenge and entertain you, then I promise to keep our lives interesting and fun. Plus, I've been told I'm quite the catch in the romance department.

Profile B: *Username:* GentlemanRon

Tagline: Let's share stories and dreams over coffee or under the stars.

About Me: If you're looking for someone who values deep, thoughtful conversations and truly cherishes the moments spent together, then we might just be a match made in heaven. I consider myself a sensitive soul, finely attuned to the subtleties of companionship, and I pride myself on being the kind of partner who listens intently and speaks words that resonate with warmth and understanding.

What I'm Looking For: I'm attracted to a woman who appreciates sincerity and heartfelt conversations. If you find joy in discussing everything from classic literature to your hopes and future dreams, and you love sharing quiet moments just as much as the lively ones, you'll find in me a willing and enthusiastic partner.

Hobbies:

- Exploring the world of literature and philosophy.
- Crafting poems that capture the essence of our daily lives.
- Designing cozy, inviting spaces in my home for long chats.
- Discovering new coffee shops for our next conversation.

Fun Facts:

- I have a knack for remembering dates — your birthday will be celebrated with as much enthusiasm as any major holiday!
- My playlist includes everything from Chopin to The Beatles, perfect for any mood or occasion. I'm known among my friends for giving thoughtful, personalized gifts that speak to the heart.

Why You Should Swipe Right: If you desire a relationship where you feel deeply understood and genuinely cherished, where conversations flow freely and connections grow strong, I am ready to offer that and more. I know when to step back and listen, and when to step up and reassure. With me, you'll never doubt that you are the focus of my attention and affection.

On MatchRate, dating profiles receive integer ratings from members of the opposite sex. These span from [0,10] with the occasional enthusiastic "11" or maybe even a "12". They follow a Poisson distribution, with a prior-posterior that is a Gamma distribution.

Ron is not concerned with the *average* count of interested ladies that would respond to him given one or the other of *profile A* or *profile B*. Ron prefers quality over quantity. Algorithmic speaking, Ron is not interested in means and other point estimates, the stuff of traditional frequentist *A/B* decision models. Rather he is interested in dating a perfect "10" — or at least, as we will find out below, a "5" or above. In statistical parlance, he is interested in the distribution of *extreme values* of his match distribution. In this sense, his decision problem is like an insurance problem dealing with extreme situations, which in an insurance context are called "risks".

Ron has decided to approach the dating game algorithmically and experiments with each of his two profiles. He collects 1000 observations for each profile, showing how the ladies rate him given a particular profile. Additionally, for each rating he receives, he also rates those giving him that rating. Figure 7.3 shows the distributions of ratings in his experimental observations.

Ron is not particularly confident in his ability to craft a dating profile, so *a priori* believes that the average rating either of his profiles will receive from the ladies will only be around a 1, which corresponds to a Poisson's $\lambda = 1$. He might do better, but he opts for a conservative prior. He constructs a prior distribution which is $Gamma(shape = 1, rate = \sqrt{1} = 1)$ from the models in Chapter 3 and Bayesian revises these with the data set that he collected from experiments with his profiles. Figure 7.4 shows his posteriors.

Let's now find out how many nights of swiping right that it will take for Ron to actually meet his match, and find a date. Let's assume that Ron and one of the ladies on MatchRate connect only when the ladies, giving Ron a significant amount of latitude as well as the benefit of the doubt, swipe right for Ron only when they rate him higher than a "5". In contrast, Ron is picky, and only responds to ladies he ranks greater than "8", i.e., matches only occur for the event $Ron(rating > 5) \cap Lady(rating > 8)$. The number of nights that Ron will need to revisit the bar scene, swiping on MatchRate,

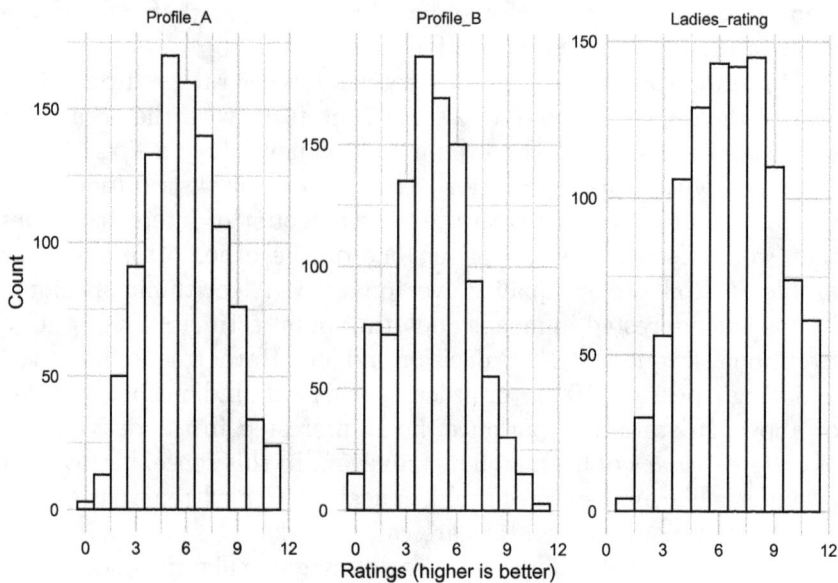

Figure 7.3. Data collection for the rating game.

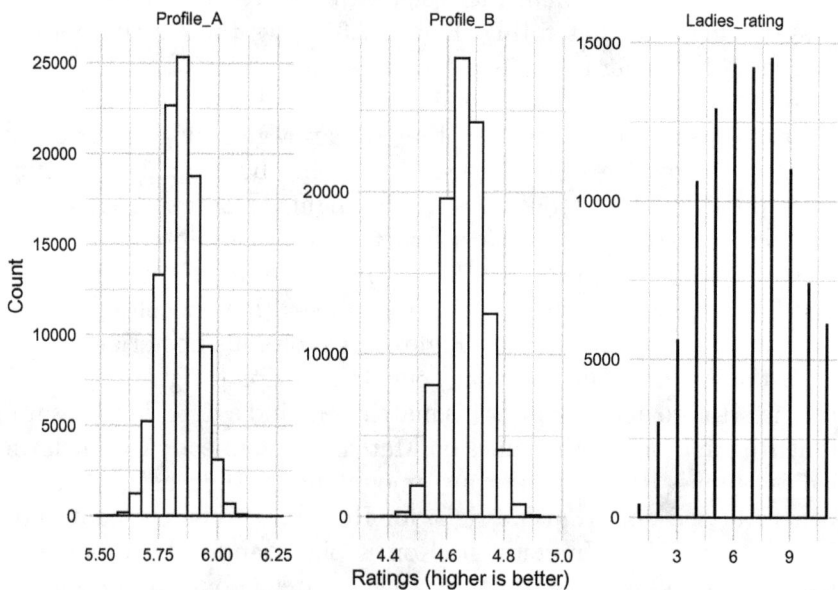

Figure 7.4. Posteriors from Bayesian revision with Ron's experiments.

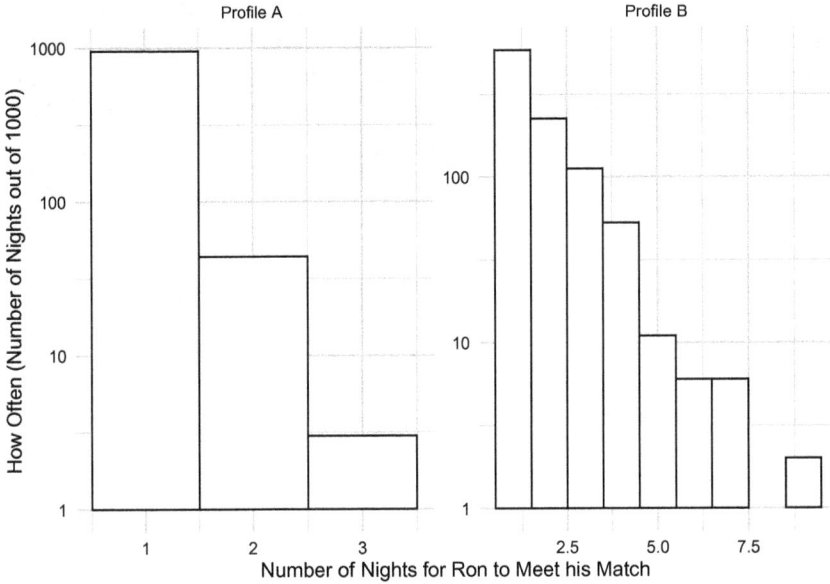

Figure 7.5. How many nights will it take Ron to find a date?

is shown in Figure 7.5 for Ron's profiles A and B. Clearly the "sexy son" profile A is much more effective (and quicker) in finding Ron a match.

For small insurance-like problems such as this, computer programming is an efficient and effective way to analyze the data and decide. But we could have achieved the same goal analytically.

For a Gamma distributed random variable, the maximum value for large sample sizes converges to one of the three types of extreme value distribution. For Gamma distributions, which do not have an upper bound and decrease exponentially, the Gumbel distribution is the appropriate limit. The maximum values from a sample of X_n Gamma distributed random variables can be approximated by rescaling to fit a Gumbel distribution as $n \to \infty$ with cumulative distribution function $G(x) = \exp(-\exp(-\frac{x-\mu}{\sigma}))$ where μ and σ are the sample mean and standard deviation of the original Gamma distribution.

We know from Chapter 3 on priors how to compute the mean and variance for Gamma functions, so we can use these values to rescale to fit the Gumbel. The R language has a number of packages to support extreme value distributions. Perhaps the most useful is the **evd** package. This has functions to support the Gumbel distribution, as well as order statistics, and just the maximum or minimum order statistics with the **extreme** function. You, the reader, can start with the R code in the Technical Appendix at the end of this book to explore further ways to model these types of real world matching problems.

Chapter 8

Bayesian A/B Decision Models for Sentiment Analysis on Textual Databases

A/B decision models are not a natural first-choice for analyzing sentiments and textual databases because the data is not numerical. Using quantitative decision models demands some sort of translation from the qualitative source. This chapter shows how, using natural language processing (NLP) and bag-of-words (BoW) methods to interpret and quantify text data, A/B tests can be applied to such data.

These models facilitate the comparison of different versions of textual data processing techniques to determine which yields the most accurate and actionable insights.

8.1. Application Areas for A/B Decision Models in Natural Language Processing

(1) **Customer Feedback Analysis:** One of the primary applications of A/B decision models in sentiment analysis is analyzing customer feedback. Businesses often receive vast amounts of unstructured feedback through surveys, reviews, and social media. A/B testing can be used to compare different sentiment analysis algorithms to determine which most accurately captures

customer sentiment, helping companies improve products and services based on customer opinions.

(2) **Social Media Monitoring:** Social media platforms generate an enormous volume of text data, making them a rich source for sentiment analysis. A/B decision models can help compare different text processing and sentiment analysis techniques to monitor brand perception, track public opinion on various topics, and identify emerging trends. This enables businesses to respond swiftly to negative sentiment and capitalize on positive trends.

(3) **Market Research:** In market research, sentiment analysis is used to gauge consumer sentiment towards brands, products, and industry trends. A/B decision models can be employed to evaluate the effectiveness of different sentiment analysis models in interpreting market research data, ensuring that the insights derived are accurate and reflective of true consumer sentiment.

(4) **Financial Sentiment Analysis:** Financial analysts use sentiment analysis to interpret news articles, earnings reports, and social media commentary to predict stock market trends. A/B testing can be used to compare different sentiment analysis techniques to identify which models best correlate with market movements, aiding in more informed investment decisions.

(5) **Content Personalization:** Personalizing content based on user sentiment can significantly enhance user engagement and satisfaction. A/B decision models can help in testing various sentiment analysis approaches to tailor content recommendations, email campaigns, and customer interactions based on the detected sentiment, ensuring that the content resonates well with the audience.

8.2. Advantages Offered by Bayesian A/B Decision Models

Bayesian A/B decision models offer significant advantages over their frequentist counterparts in three areas: optimization, user experience, operating costs and risk.

(1) **Accuracy and Replicability of Sentiment Analysis:** A/B decision models address the challenge of accuracy in sentiment

analysis by enabling the comparison of different models and techniques. By testing variations, companies can identify which methods produce the most accurate sentiment scores, leading to better decision-making and more reliable insights.

(2) **Handling Unstructured Data:** Sentiment analysis typically deals with unstructured text data, which can be challenging to analyze. A/B testing helps in evaluating different text processing techniques, such as tokenization, stemming, and lemmatization, to determine which approaches most effectively prepare text data for sentiment analysis.

(3) **Model Selection:** With numerous sentiment analysis models available, from simple lexicon-based approaches to complex machine learning models, A/B decision models assist in selecting the best-performing model. This ensures that the chosen model provides the highest accuracy and reliability for the specific application context.

8.3. Caveats and Pitfalls in Bayesian Approaches

The analyst must keep aware of specific problems involved in Bayesian A/B tests, many of which are also shared in one form or another by frequentist tests.

(1) **Quantitative Analysis of Qualitative Data:** One of the unique challenges of applying A/B decision models to sentiment analysis is the difficulty of conducting quantitative analysis on qualitative text and sentiment data. Textual data is inherently subjective and context-dependent, making it challenging to translate into numerical values that can be compared objectively.

(2) **Ambiguity and Context:** Sentiment analysis often struggles with ambiguity and context, as words can have different meanings in different contexts. Sarcasm, irony, and cultural nuances further complicate the analysis. A/B testing might reveal that a model performs well in one context but poorly in another, highlighting the importance of context-aware sentiment analysis.

(3) **Data Preprocessing Challenges:** The preprocessing stage is crucial for sentiment analysis, involving tasks such as text cleaning, tokenization, and normalization. Different preprocessing techniques can significantly impact the outcome of the sentiment

analysis, and A/B testing must account for these variations to ensure fair comparisons.

(4) **Selection of Sentiment Labels:** Sentiment analysis models typically classify text into predefined categories such as positive, negative, or neutral. However, the granularity of these categories can vary. A/B testing different granularity levels (e.g., binary vs. multi-class sentiment labels) can lead to differing conclusions about the superior model.

(5) **Overfitting and Generalization:** Overfitting is a common pitfall in sentiment analysis, where a model performs exceptionally well on training data but fails to generalize to new data. A/B testing must ensure that models are evaluated on diverse and representative datasets to avoid overfitting and ensure robustness.

(6) **Computational Complexity:** Advanced sentiment analysis models, especially those based on deep learning, can be computationally intensive. A/B testing these models requires significant computational resources, which can be a constraint for many organizations.

Bayesian A/B decision models with innovative problem setups can, and do play a critical role in optimizing sentiment analysis techniques for textual databases, offering a structured approach to comparing different models and methods. Such models offer a significant extension of the A/B decision models traditionally used in marketing, because they can actually listen to the voice of customers, to find out what advertising and marketing campaigns work, and which ones are ignored or despised by consumers. These models help address key challenges in sentiment analysis, such as accuracy, handling unstructured data, and model selection. However, the unique difficulty of conducting quantitative analysis on qualitative text and sentiment data presents significant challenges.

The inherent subjectivity and context-dependence of textual data make it challenging to derive reliable numerical metrics for A/B testing. Ambiguity, sarcasm, and cultural nuances further complicate sentiment analysis, requiring context-aware approaches. Data preprocessing techniques and the selection of sentiment labels also significantly impact sentiment analysis outcomes, necessitating careful consideration during A/B testing.

Moreover, the risk of overfitting and the need for significant computational resources highlight the importance of robust model evaluation and efficient resource management. Despite these challenges, A/B decision models remain a powerful tool for refining sentiment analysis techniques, enabling organizations to derive more accurate and actionable insights from textual data.

By navigating these caveats and pitfalls thoughtfully and strategically, businesses can leverage A/B decision models to enhance their sentiment analysis capabilities, leading to better customer insights, more effective marketing strategies, and improved decision-making. As the field of natural language processing continues to evolve, the application of A/B testing in sentiment analysis will undoubtedly play a crucial role in advancing the accuracy and effectiveness of text-based analytics.

8.4. Why Don't We Use Large Language AI Models for A/B Tests?

We often would like to determine whether, in a pair of documents, how the sentiment of the two documents compares. Is one text more positive, more liberal, more loved, angrier, sadder, disgusting, fearful, joyful, or trustworthy? These concepts, though eminently meaningful to humans, and not easily translated into mathematics.

Large language models (LLMs) use deep learning to understand and generate language. They are trained on large amounts of text data using a self-supervised and semi-supervised process that involves learning statistical relationships from text documents. They can perform a variety of tasks, including recognizing, summarizing, translating, predicting, and generating text, and performing sentiment analysis.

Impressive as they are, LLMs tend not to be ideal for statistical A/B decisions. Human language involves a great deal of uncertainty and variability in expression. To sound more natural, LLMs have simulated this variability and allowed it to be tuned via a "temperature of discourse" parameter. Thus, successive prompts of an LLM will generate different outputs. This variability causes obvious problems in A/B decisions: the first analysis might decide that choice A is the better choice; but next time the analysis is performed, the choice would be B. LLM modeling is not replicable, and that undermines its usefulness for statistical tests.

8.5. The Bag-of-Words Model

Replicability demands a simpler model for natural language processing (NLP). The bag-of-words (BoW) model is a foundational approach in text processing, particularly utilized in NLP and information retrieval (IR). It represents text as an unordered collection, or "bag", of words, disregarding word order and grammar, yet capturing word multiplicity. The BoW model is frequently employed in document classification methods. Here, the frequency of each word's occurrence is used as a feature for training classifiers.

Consider two simple text documents:

`Fred likes to watch NASCAR races; Masha likes NASCAR too.`

`Masha also likes to watch karate competitions.`

For each document, a list is constructed based on the words present:

Document 1: "Fred", "likes", "to", "watch", "NASCAR", "Masha", "likes", "NASCAR", "too"

Document 2: "Masha", "also", "likes", "to", "watch", "karate", "competitions"

Each bag-of-words can be represented as a JSON object:

BoW1 = ["Fred": 1, "likes": 2, "to": 1, "watch": 1, "NASCAR": 2, "Masha": 1, "too": 1]

BoW2 = ["Masha": 1, "also": 1, "likes": 1, "to": 1, "watch": 1, "karate": 1, "competitions": 1]

Each key represents a word, and each value represents the number of occurrences of that word in the document. In the context of bag algebra, the "union" of two documents in the BoW representation is the disjoint union, summing the multiplicities of each element $BoW3 = BoW1 \uplus BoW2$.

Implementations of the BoW model often use word frequencies to represent a document's contents. These frequencies can be normalized by the inverse of document frequency, or tf–idf. Additionally, for classification purposes, supervised alternatives have been developed to account for the class label of a document. Binary weighting (presence/absence or 1/0) is also employed for certain problems, such as those implemented in the WEKA machine-learning software system.

8.6. Case Study: A/B Testing in Product Design at Madcap Macaroons

8.6.1 *Introduction*

Madcap Macaroons is a burgeoning cookie company based on the West Coast, famous for popular treats like Banshee Biscuits, Sinister Snaps, Diabolical Delights, and Cackle Cookies. Madcap was facing a pivotal challenge in one particular product's design. With limited budget and resources, the company was at a crossroads, needing to decide between two competing designs for their flagship cookies This case study explores how Madcap leveraged A/B testing to make an informed product design choice that would ultimately resonate with their target market.

8.6.2 *Design dilemma*

Madcap Macaroons were famous for their unique shapes enhanced with proprietary flavor and scent technologies, and their proprietary online purchase platform with home deliveries only. The company's founders argued that a complete sensory experience should not limit users just to sugar rush, but should include tastes and scents to titillate the palate and provide an ultimate immersive massage experience. Yet the addition of these new cookie dimensions spawned countless debates within the company on application and design, and just what foods and perfumes would best enhance the user's overall gustatory experience. The company was enthused about branding opportunities, particularly with fast food restaurants and superstores, but walked a fine line between innovation, fine art and kitschy imitation bordering on the eccentric. Their engineers and food chemists, though brilliant in their own realms, were sadly not the most reliable arbiters of consumer taste.

The engineering team at Madcap Macaroons was divided into two factions, each championing a different aesthetic based on distinct consumer experiences:

- **Design A:** Joker's Grins featured a soft, edible center, which was particularly suited for use with frostings and provided a comforting sensation and mouthfeel, but sadly was prone to dripping on clothes.
- **Design B:** Mischief Munchies boasted a crunchy center and spongy surface, with a more crackling sensation and mouthfeel, suitable for neat, on-the-go snacking.

Both teams had conducted extensive laboratory tests in their customized laboratories with numerous volunteers and family members. Each was emphatic about the superiority of their cookies in satisfaction, hygiene and gustatory superiority.

Given the constraints of their budget, Madcap Macaroons could only afford to move forward with one of these designs.

8.6.3 *Methodology*

To resolve this dilemma, Madcap Macaroons embarked on an innovative A/B testing approach using limited production runs. They distributed both designs through their online sales platforms, allowing real-world customer interactions to guide their final decision.

The primary metrics for evaluation were customer reviews and product returns. Reviews were collected from various online platforms, featuring a 1 to 5 star rating system, along with detailed textual feedback in the review titles and content. Along with each review, the system captured the total sales to that customer.

8.6.4 *Data collection and analysis*

The test marketing period spanned one month, during which both massagers were sold simultaneously on identical platforms and markets. This setup ensured a fair comparison of customer preferences.

Post-sales, the reviews for each product were aggregated and analyzed using advanced NLP techniques. The NLP analysis employed the "nrc" lexicon, which categorized words into a set of binary sentiments: *positive, negative, anger, anticipation, disgust, fear, joy, sadness, surprise,* and *trust.* This method converted qualitative text into quantifiable sentiment scores, providing a nuanced understanding of customer emotions and perceptions.

8.6.5 *Results*

The A/B testing revealed clear preferences in customer satisfaction and sentiment, demonstrating a marked difference in how each design was received by the target audience. The detailed sentiment analysis offered deep insights into the emotional responses elicited by each design, which were instrumental in making the final decision.

Initially they "eyeballed" the data using word clouds (Figures 8.1 and 8.2) to obtain a sense of the overall sentiment towards each of their recipes.

MadCap Cookies gathered 584,454 reviews with a total of 785,087 words. They removed stop words, i.e., short words such as a, that and so forth that don't convey sentiment and distilled the reviews into 120,738 words to which their lexicon attached a particular sentiment, such as joy, surprise, trust and so forth (Figures 8.3 and 8.4).

Initial investigations and data analysis showed the following contributions from the most influential review words for each of the choices.

Figure 8.1. Word cloud for choice A: Joker's Grins.

Figure 8.2. Word cloud to sentiments for choice B: Mischief munchies.

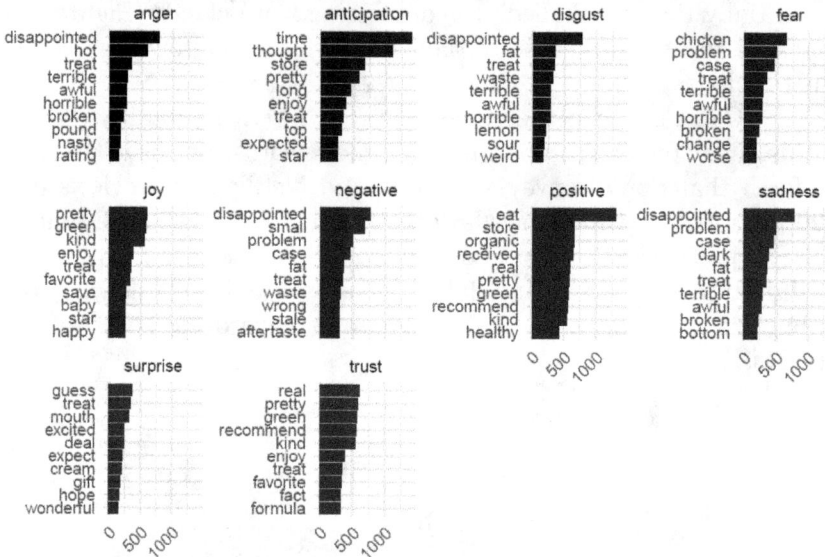

Figure 8.3. Contributions to sentiments for choice A.

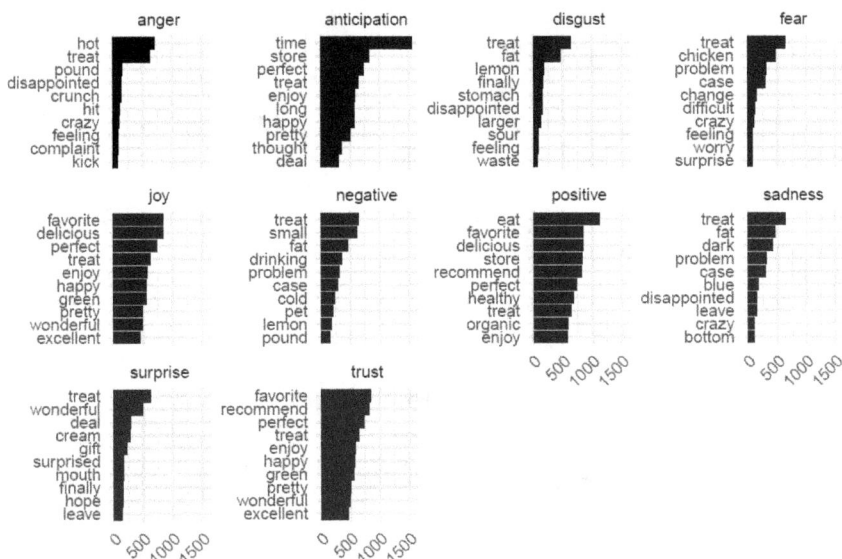

Figure 8.4. Contributions to sentiments for choice B.

Madcap summarized these reviews in terms of posterior means, which gave a clearer picture of reviewers' sentiments toward the two cookie recipes (Figures 8.5–8.14).

Sentiment	Mean for choice A (Joker Grins)	Mean for choice B (Mischief munchies)
Anger	21.2	14.7
Anticipation	47.2	46.5
Disgust	24.9	15.8
Fear	18.9	12.5
Joy	59.0	66.8
Negative	18.5	12.8
Positive	38.0	39.6
Sadness	24.6	16.8
Surprise	41.3	42.8
Trust	42.9	44.8

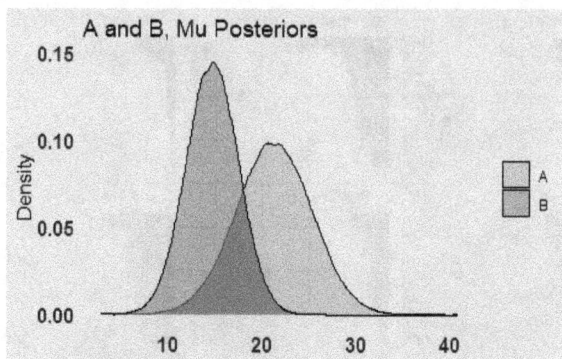

Figure 8.5. Anger: *A* (Joker's Grins) versus *B* (Mischief Munchies) outcome on *AB* test.

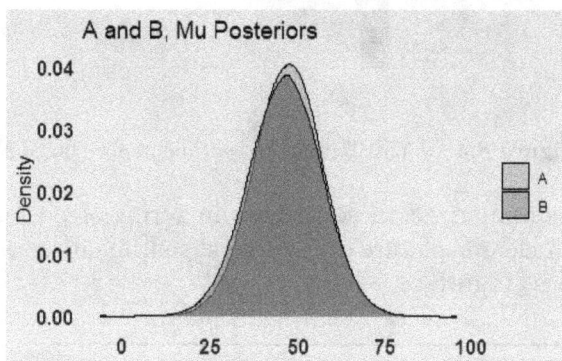

Figure 8.6. Anticipation: *A* (Joker's Grins) versus *B* (Mischief Munchies) outcome on *AB* test.

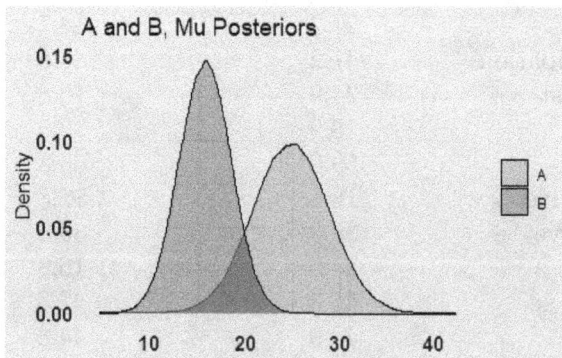

Figure 8.7. Disgust: *A* (Joker's Grins) versus *B* (Mischief Munchies) outcome on *AB* test.

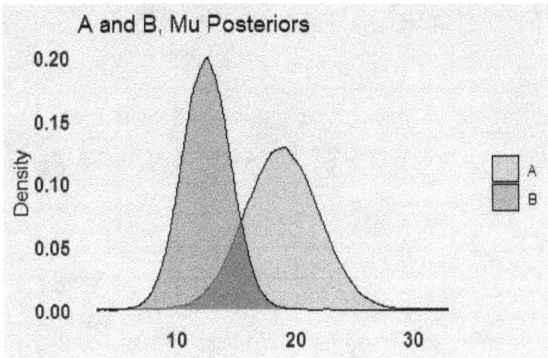

Figure 8.8. Fear: *A* (Joker's Grins) versus *B* (Mischief Munchies) outcome on *AB* test.

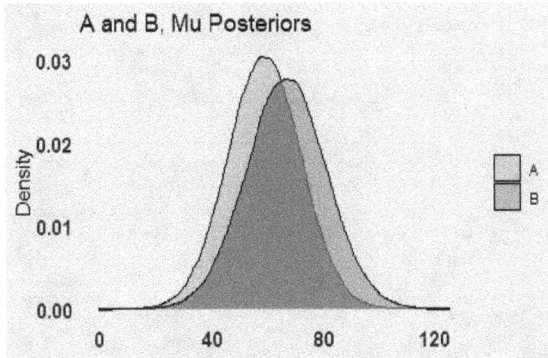

Figure 8.9. Joy: *A* (Joker's Grins) versus *B* (Mischief Munchies) outcome on *AB* test.

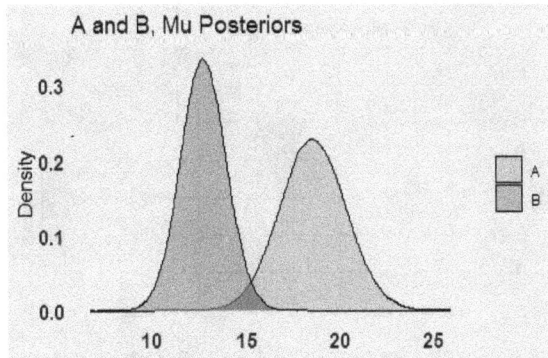

Figure 8.10. Negative: *A* (Joker's Grins) versus *B* (Mischief Munchies) outcome on *AB* test.

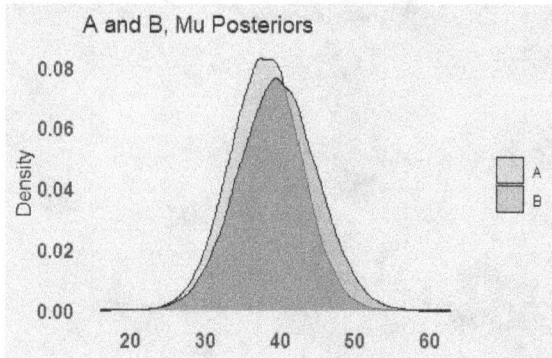

Figure 8.11.   Positive: A (Joker's Grins) versus B (Mischief Munchies) outcome on AB test.

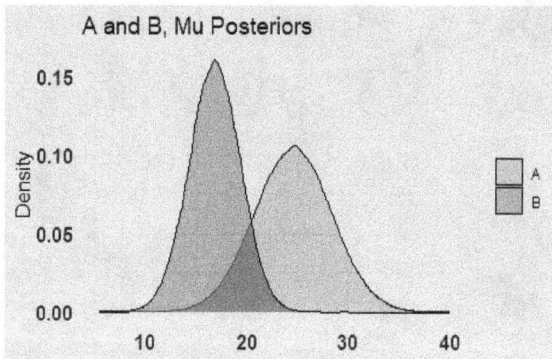

Figure 8.12.   Sadness: A (Joker's Grins) versus B (Mischief Munchies) outcome on AB test.

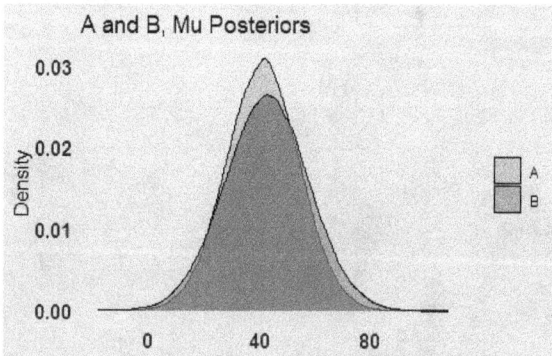

Figure 8.13.   Surprise: A (Joker's Grins) versus B (Mischief Munchies) outcome on AB test.

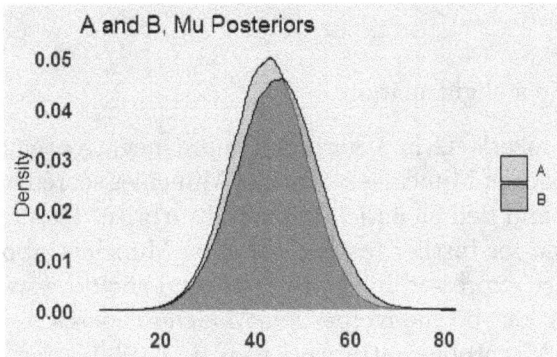

Figure 8.14. Trust: A (Joker's Grins) versus B (Mischief Munchies) outcome on AB test.

Graphical A/B posteriors provide perspective on the variance of the responses, which could be seen as a measure of the certainty or equivocality of reviewers' sentiments.

Madcap considered themselves data agnostic, and felt it was only fair to each of the teams to invoke diffuse priors for the tests. They assumed Normally distributed data for their Bayesian revision.

They then summarized two data sets: one for choice A (Joker's Grins) with 6,066 words and one for choice B (Mischief Munchies) with 5,709 words. Finally they summarized around sentiments into choice A and choice B lists of two statistics associated with each of the 10 sentiments in the **nrc** lexicon: the mean number of words associated with that sentiment, and the standard deviation in word counts associated with that sentiment.

8.6.6 *Conclusion*

The results of A/B testing across the sentiments gave the following summary choices between A and B:

- anger: A
- anticipation: *tie*
- disgust: A
- fear: A
- joy: B by a slight margin
- negative: A
- positive: B by a slight margin

- sadness: A
- surprise: *tie*
- trust: B by a slight margin

Choice A (Joker's Grins) garnered more negative emotions than choice B (Mischief Munchies); Mischief Munchies scored well on trust, joy, positive and tied on anticipation and surprise. Though there may be some room for further testing, Mischief Munchies appeared to be the preference expressed in the reviews, and should, given the available information, be the recipe that Madcap chooses.

Madcap Macaroons' strategic use of A/B testing in product design not only resolved the internal debate but also enhanced their understanding of the customer's needs and preferences. This case study exemplifies the power of combining traditional marketing strategies with cutting-edge data analysis techniques to make informed decisions in product development. The success of this approach highlights the potential of A/B testing as an essential tool for product designers seeking to match market demands and maximize customer satisfaction.

8.6.7 *Thoughts on using informative priors*

The Madcap data analysts considered themselves agnostic, and chose to apply noninformative priors in the case study. But when decisions such as these need to be made on a recurring basis, data analysts are likely to develop intuitions over time, and these intuitions are potentially a basis for constructing informative priors. In the current case study, data was assumed to be Normally distributed. How might an informative prior be constructed from existing knowledge about the firm's business and customers?

The interpretation of parameters in the Normal-inverse-gamma distribution is the conjugate prior of a Normal distribution with unknown mean and variance. It is the Normal-inverse-gamma distribution but uses the precision (common with Bayesians; precision is the inverse of variance, and makes many of their calculations simpler) rather than the variance.

The parameters may be loosely estimated as follows:

- μ is the estimate of the mean from your subjective assessment of beliefs and other sources of information.

- $\lambda \in [0, \infty)$ is a measure of your confidence in the accuracy of the mean estimate μ that you provided based on your subjective assessment of beliefs and other sources of information. Generally you would like to keep this small, indicating that you are careful not to overweight the prior in the analysis.
- $\alpha \in [0, \infty)$ is a measure of your confidence in the accuracy of the standard deviation σ that you have provided based on your subjective assessment of beliefs and other sources of information. Generally you would like to keep this small, indicating that you are careful not to overweight the prior in the analysis.
- $\beta = \frac{\alpha \sigma^2}{2}$ is the estimate of the scaled variance of the observations used in computing the likelihood function. In this calculation σ is the estimate of the standard deviation from your subjective assessment of beliefs and other sources of information.

Let's assume for our example that we have gathered opinions from focus groups about each of the options and have collated all of their responses into the following prior parameter settings for Bayesian A/B estimation using the rules of thumb above. For priors, we can ignore details required for statistically accurate setting of these parameters, and these approximations will be sufficient because in Bayesian revision, they will be updated by the data.

8.6.8 *Likert scale surveys for subjective stated preferences*

Another common way that marketers gather consumer preference data is through online surveys gathering consumer responses to questions or statements in 5 or 7-point Likert scales. Tools like SurveyMonkey have made it effortless to prepare and conduct surveys. This too often allows lazy and untrained researchers to construct poorly designed questionnaires, collect data that is highly biased, error-ridden and unreliable and apply inappropriate methods to draw conclusions. Such surveys are worthless, but the use of computer software to gather and statistically analyze the data will provide the results with a patina of respectability, their conclusions being presented as credible. Even where survey methods are applied flawlessly, the data itself does not represent true consumer preferences; rather, it captures the "stated" preferences of consumers who may lie, mislead, or not pay attention to the survey questions. True consumer preferences are found in purchases and after-purchase reviews of products — what are called "revealed" preferences. The only way to obtain these true-revealed preferences is to obtain consumers' actual purchase decisions. Nonetheless, survey data can provide useful information if they are instead interpreted as subjective priors — subjective information to be updated by more reliable revealed preferences in consumer purchase and review data. Bayesian methods provide roles for both stated (subjective) and revealed (objective) consumer preferences. [36] showed that Likert scaled data is well-modeled by a Normal distribution despite data being truncated at zero (i.e., there are no negative Likert responses).

The digital age has revolutionized how marketers understand consumer preferences. With platforms like SurveyMonkey, creating and conducting online surveys has become a breeze! These tools empower marketers to quickly gather consumer responses using 5 or 7-point Likert scales, making the collection of consumer sentiment data more accessible than ever.

However, with great power comes great responsibility. The ease of creating surveys has sometimes led to less rigorous research practices. Occasionally, this results in surveys that are poorly designed, biased, or error-prone, which can mislead conclusions despite the sheen of sophistication that statistical software might lend to the results.

Traditional survey methods capture what we call "stated" preferences — what people say they prefer, rather than what they actually do. But there's a catch! These responses might not always reflect their true intentions, as factors like social desirability or lack of attention during the survey can skew results. So, how do we get to the heart of what consumers really want?

The answer lies in "revealed preferences" — the powerful insights we gain from actual consumer behaviors, such as purchases and product reviews. This data is gold, showing us not just what consumers say, but what they actually do. Understanding these true preferences is a game-changer for product design and marketing strategies.

The integration of Bayesian methods into market research offers an exciting frontier for harnessing the full spectrum of consumer insights. These methods allow us to treat survey data as "subjective priors" — initial assumptions that can be updated with solid, objective data gleaned from actual consumer actions. This dynamic approach enables a richer, more accurate understanding of the market, blending stated and revealed preferences in a way that respects both the subtleties of human behavior and the realities of consumer actions.

As we move forward, the ability to merge traditional survey data with empirical consumer behavior opens up thrilling possibilities. Marketers are now equipped to draw more nuanced, actionable insights that can drive smarter, more consumer-aligned product developments. The future of consumer preference analysis is here, and it's vibrant, insightful, and incredibly promising!

Chapter 9

Legal Applications of A/B Decision Models

9.1. Bayesian A/B as an Alternative Model for Event Studies

Event studies investigate financial metrics before and after a particular event or class of events has happened in order to determine the magnitude of influence of that event on financial performance. The event may occur at a specific point in time, such as a stock market crash. It may also occur at a multitude of differing times, in which case stock prices or corporate earnings will be synchronized around the event time on their own time series. For example, the impact of a corporation's switch from LIFO accounting to FIFO accounting would occur at unique times for different corporations, but their earnings time series could be synchronized around the time of the switch to determine the impact in general on earnings in an industry or economy-wide. The purpose of the event study, again, is to determine the impact and magnitude of classes of events on the finances and economics of corporations.

Dolley [37] published seminal work in the price effects of stock splits. Subsequent authors [38–42] improved event study methodologies by removing general stock market price movements and separating out confounding events. Ball and Brown [43] and Fama [44,45] introduced the methodology that is in use today. Event studies are used in the field of law and economics to measure the impact on the value of a firm of a change in the regulatory environment [46].

In legal liability cases, event studies have been used to assess damages [47]. In the majority of applications, the focus is on the effect of an event on the price of a particular class of securities of the firm, most often common equity. The methodology sometimes substitutes streams of Hicksian demands [48] derived from Google search and advertising data for the traditional streams of financial exchange-reported asset prices used in financial event studies.

In the years since these pioneering event studies, a number of modifications have been developed. These modifications relate to complications arising from violations of the statistical assumptions used in the early work and relate to adjustments in the design to accommodate more specific hypotheses [43,49–51]. MacKinlay [52] observes that there is no unique event study structure; rather, there is a general flow of analysis. The initial task of conducting an event study is to define the event of interest and identify the period over which market prices of the firms involved in this event will be examined, that is, the event window. For example, if one is looking at the information content of earnings with daily data, the event will be the earnings announcement and the event window will include the one day of the announcement. It is customary to define the event window to be larger than the specific period of interest. In practice, the period of interest is often expanded to multiple days or weeks. The periods prior to and after the event may also be of interest. For example, in the earnings announcement case, the market may acquire information about the earnings prior to the actual announcement, and one can investigate this possibility by examining estimation window returns. After identifying the event, it is necessary to determine the selection criteria for the inclusion of a given firm in the study, the length of the time-series, and to discuss any potential biases that may have been introduced through the sample selection. Short-horizon event studies are more reliable than long-horizon event studies as the latter have many limitations; Warner [49] describes methodologies to improve the design and reliability of the studies over longer periods.

Event studies have been staples of legal damage calculations and supporting calculations for other financial aspects of legal cases for almost a century. Yet there has yet to be a consensus even today concerning the best mathematics to ensure comparability and interpretability of results. As the following case study will demonstrate, event studies can be performed using Bayesian A/B tests. The results with Bayesian A/B tests are superior to prior frequentist approaches to analyzing the impact of events on financial metrics because the entire posterior distribution is available and can be used to opine on risk, return, and other important characteristics of an earnings stream.

9.2. Applications of A/B Testing in Law

Decision models for A/B testing have many potential applications in commercial law, torts, and criminal law. A/B testing can support damages calculations, evidence evaluation, and the determination of guilt and culpability, and can enhance the accuracy and reliability of legal processes.

In commercial law, disputes often arise over contract breaches, intellectual property issues, and other business-related conflicts. A/B testing can be instrumental in calculating damages, evaluating the impact of disputed actions, and providing evidence to support legal arguments.

(1) **Damages Calculation:** Calculating damages accurately is crucial in commercial disputes. A/B testing can help by isolating the effects of a specific action or breach on business performance:

    (a) *Lost Profits:* In cases where a breach of contract or unfair competition leads to lost profits, A/B testing can compare the performance of the business before and after the breach. By creating a control group (pre-breach data) and a test group (post-breach data), it is possible to estimate the impact of the breach on revenue and profits. This method provides a data-driven approach to quantify lost profits, making the damages calculation more precise and defensible in court.

    (b) *Market Impact:* For intellectual property disputes, such as trademark infringement or patent violations, A/B testing can assess the market impact of the infringement. By comparing consumer behavior with and without exposure to the infringing product, it is possible to determine the extent to which the infringement affected market share, brand perception, and sales. This evidence can be used to support claims for damages or injunctive relief.

(2) **Evidence Evaluation:** A/B testing can also support the evaluation of evidence in commercial law cases, helping to establish causation and the effectiveness of remedies:

   (a) *Causation Analysis:* Establishing causation is critical in many commercial disputes. A/B testing can help demonstrate a causal relationship between a defendant's actions and the plaintiff's losses. For example, in false advertising cases, A/B testing can compare consumer responses to truthful versus misleading advertisements, providing evidence of how the misleading ads influenced purchasing decisions.

   (b) *Remedy Effectiveness:* When determining appropriate remedies, such as corrective advertising or changes to business practices, A/B testing can evaluate the effectiveness of proposed measures. By testing different versions of corrective actions, it is possible to identify the most effective remedy, ensuring that the chosen solution addresses the issue and mitigates further harm.

In tort law, A/B testing can be used to support claims for personal injury, product liability, and negligence. It can help calculate damages, evaluate the impact of tortious actions, and provide evidence for causation and liability.

(1) **Damages Calculation:** A/B testing can assist in calculating damages in tort cases by comparing the injured party's condition or circumstances before and after the tortious act:

   (a) *Medical Expenses:* For personal injury claims, A/B testing can compare medical outcomes with and without specific treatments or interventions. By analyzing data from patients who received different treatments, it is possible to estimate the cost and effectiveness of medical care required due to the injury. This information can support claims for medical expenses and future care costs.

   (b) *Loss of Earnings:* In cases involving loss of earnings, A/B testing can compare the plaintiff's income trajectory before and after the injury. By creating a control group of individuals with similar career paths and comparing their earnings growth with the plaintiff's post-injury earnings, it is possible to estimate the lost earnings and future earning potential. This approach provides a robust basis for calculating economic damages.

(2) **Evidence Evaluation:** A/B testing can provide evidence to support claims of causation and liability in tort cases, helping to establish the connection between the defendant's actions and the plaintiff's harm.

   (a) *Product Liability:* In product liability cases, A/B testing can evaluate the safety and performance of products under different conditions. By comparing data from users of the defective product (test group) and users of a non-defective version (control group), it is possible to demonstrate the defect's impact on safety and performance. This evidence can support claims for design or manufacturing defects and help establish liability.

   (b) *Negligence:* For negligence claims, A/B testing can assess the effectiveness of safety measures or procedures. By testing different safety protocols, it is possible to determine which measures could have prevented the injury or harm. This evidence can be used to argue that the defendant's failure to implement effective measures constituted negligence.

In criminal law, A/B testing can support the determination of guilt and culpability, providing evidence to establish causation, intent, and the effectiveness of interventions:

(1) **Evidence Evaluation:** A/B testing can help evaluate evidence in criminal cases, particularly in establishing causation and the impact of the defendant's actions.

   (a) *Forensic Analysis:* A/B testing can be used in forensic analysis to compare different scenarios and determine the most likely cause of an event. For example, in arson investigations, A/B testing can compare fire patterns from different ignition sources to identify the cause of the fire. This evidence can help establish whether the fire was accidental or intentional.

   (b) *Behavioral Analysis:* In cases involving behavioral evidence, such as fraud or cybercrime, A/B testing can analyze user behavior under different conditions. By comparing data from users exposed to fraudulent schemes (test group) and those not exposed (control group), it is possible to demonstrate how the fraudulent actions influenced behavior and caused harm.

(2) **Determination of Guilt and Culpability:** A/B testing can support the determination of guilt and culpability by providing evidence of the defendant's intent and the effectiveness of interventions:

(a) *Intent and Motive:* Establishing intent and motive is crucial in criminal cases. A/B testing can analyze patterns of behavior that suggest intent. For example, in fraud cases, A/B testing can compare the defendant's transactions before and after the alleged fraud to identify patterns indicative of deliberate deception. This evidence can help establish the defendant's intent and support the prosecution's case.

(b) *Effectiveness of Interventions:* A/B testing can evaluate the effectiveness of interventions aimed at preventing criminal behavior. For example, in rehabilitation programs for offenders, A/B testing can compare recidivism rates between participants in different types of programs (control and test groups). This evidence can help determine which interventions are most effective in reducing reoffending and inform sentencing decisions.

9.3. Caveats

Analysts should be keenly aware that legal applications of A/B testing are subject to a high standards of conduct. While A/B testing offers significant potential in supporting legal proceedings, it is essential to consider ethical issues and limitations.

(1) **Informed Consent:** In legal contexts, obtaining informed consent from participants in A/B tests can be challenging. In some cases, it may not be possible to inform individuals that they are part of a test without compromising the study's validity. Legal professionals must balance the need for accurate data with ethical considerations and ensure that participants' rights are protected.

(2) **Data Privacy:** A/B testing often involves the collection and analysis of personal data. Legal professionals must ensure compliance with data privacy laws and regulations, such as the GDPR and CCPA. This includes obtaining consent, anonymizing data, and implementing robust security measures to protect participants' privacy.

(3) **Validity and Reliability:** The validity and reliability of A/B testing results depend on the design and implementation of the tests. Legal professionals must ensure that tests are appropriately designed, with sufficiently large sample sizes and controlled conditions, to produce accurate and reliable results. Poorly designed tests can lead to incorrect conclusions and undermine the credibility of the evidence.

(4) **Ethical Use of Evidence:** The use of A/B testing in legal proceedings raises ethical questions about the manipulation and interpretation of data. Legal professionals must ensure that evidence derived from A/B testing is presented accurately and transparently, without misleading the court or manipulating the outcomes. Ethical considerations should guide the use of A/B testing to ensure that justice is served.

9.4. Case Study: VinLuxe Innovations Sues an Intellectual Property Thief

In the verdant valleys of Napa, where the wineries stretch like green-gold ribbons through the landscape, a company named VinLuxe Innovations revolutionized the wine industry with its proprietary intellectual property (IP). For years, the science of winemaking had been bound by tradition and time, a slow dance of fermentation that could not be hurried. But VinLuxe, inspired by a blend of Silicon Valley innovation and old-world sensibility, had shattered these age-old shackles. They developed a process — a closely guarded trade secret — capable of transforming the humblest of grape juices into wines that rivaled the finest Grand Crus of France. What once took decades was now achieved in just one hour.

VinLuxe's breakthrough hinged on a sophisticated chemical synthesis and molecular realignment process, a secret so valuable it was protected under the USPTO as a trade secret. The heart of this magic was in a nondescript building, fortified like a bank vault, where canned grape juice concentrate was metamorphosed into bottles of luxurious wine that connoisseurs could scarcely believe were born from humble beginnings.

The company prospered in silence and secrecy, its directors and scientists the sole keepers of the miraculous technique. Markets were captured, and glasses were raised in toast after toast to the health of VinLuxe's enigmatic prowess. For a decade, the secret remained within the sterile, high-security labs of VinLuxe's headquarters. But such secrets are like vintage wine themselves — rare and irresistible.

In June of 2015, that secret found its way into the hands of Jasper Kildare, a hacker of extraordinary skill and dubious morals.

Jasper had infiltrated VinLuxe's security under the guise of a routine digital maintenance contractor. Over months, he wove a web of digital espionage that culminated in the heist of the century: he stole the trade secret files and several vials of the chemical catalysts used in the aging process.

With these stolen treasures, Jasper established a clandestine operation in an abandoned warehouse in Oregon. His setup was rudimentary compared to VinLuxe's, but it was effective. He began producing wines so exquisite that they began to eclipse even the reputation of VinLuxe's products. Kildare Wine, as he branded it, became the new darling of wine enthusiasts and sommeliers, its origin as murky as a foggy vineyard at dawn.

VinLuxe's suspicions were first aroused by a stray batch number noted in a wine-tasting competition, a number that corresponded to none of their production lines. Further investigation revealed the shocking extent of Jasper's operation. VinLuxe's board was furious and immediately sought legal recourse. They filed a lawsuit under the Defend Trade Secrets Act of 2016 (DTSA), claiming theft and unauthorized use of their proprietary processes.

The legal battle that ensued was as complex and layered as the wines at its heart. VinLuxe's attorneys argued fiercely, presenting evidence of the security measures breached and the direct correlations between their secret processes and the products Jasper had been selling. They claimed that Jasper's actions not only constituted a clear violation of the DTSA but also damaged their market position and diluted the prestige of their brand.

Jasper's defense was brazen. He claimed that the process, while inspired by VinLuxe's methods, had been significantly altered by his own innovations. His lawyer, a sharp-witted woman with a taste for underdog cases, argued that the chemical alterations introduced by Jasper constituted a new invention altogether.

9.4.1   *The Grand Cru-sade: Calculating the cost of betrayal*

The determination of VinLuxe to recover its losses led them to launch a calculated and meticulous legal assault aimed at Jasper's unjust self-enrichment.

The first task was to ascertain the amount of profit Jasper had made from sales of wines produced using the stolen trade secret. VinLuxe's legal team, led by the sharp and seasoned attorney Elizabeth Grant, began by securing sales records from Jasper's operations through court orders. Analysts were brought in to review these records and calculate the revenue generated from these sales. The team then estimated the profit margin based on industry standards and the operational cost data provided during discovery, giving them a clear figure of Jasper's net profits (Tables 9.1 and 9.2)

Accounting values, being positive, discrete integers and right skewed, are best modeled as Poisson distributions. Since this is a time series of annual reports, they are not all stated in equal dollars, and a price deflator needs to be applied to correct for inflation bias. Otherwise, we would be biased toward concluding that in later years, Kildare had been more profitable, even if the counterfactual were true (i.e., that he did not steal trade secrets).

Table 9.1. Kildare Wine's income statement ($000).

Year	Revenues	Cost of goods sold	Gross profit	R&D	Operating expenses	Net income
2005	200	150	50	10	30	10
2006	210	157	53	10	31	12
2007	220	165	55	10	32	13
2008	230	172	58	10	34	14
2009	240	180	60	10	36	14
2010	250	187	63	10	37	16
2011	260	195	65	10	38	17
2012	270	202	68	10	40	18
2013	280	210	70	10	42	18
2014	290	217	73	10	43	20
2015	300	225	75	10	45	20
2016	500	300	200	5	60	135
2017	550	330	220	5	66	149
2018	600	360	240	5	72	163
2019	650	390	260	5	78	177
2020	700	420	280	5	84	191
2021	750	450	300	5	90	205
2022	800	480	320	5	96	219
2023	850	510	340	5	102	233
2024	900	540	360	5	108	247

Table 9.2. Kildare Wine's balance sheet ($000).

Year	Assets	Liabilities	Owner's equity
2005	500	200	300
2006	520	208	312
2007	540	216	324
2008	560	224	336
2009	580	232	348
2010	600	240	360
2011	620	248	372
2012	640	256	384
2013	660	264	396
2014	680	272	408
2015	700	280	420
2016	1400	560	840
2017	1450	580	870
2018	1500	600	900
2019	1550	620	930
2020	1600	640	960
2021	1650	660	990
2022	1700	680	1020
2023	1750	700	1050
2024	1800	720	1080

The first analysis made by Elizabeth Grant pertained to Kildare Wine's annual net income. These skyrocketed from a negligible $10,000 annually to almost $200,000 annually — a 2000% increase. Figure 9.1 shows the difference in posterior lambda (the mean of the posterior distribution) before the alleged theft of IP and after. The earnings impact of the theft was dramatic.

To solidify their claim, they compared these figures with market analytics to demonstrate how Jasper's entry into the market with a similar product at a slightly lower price point directly siphoned off VinLuxe's existing and potential customer base (Figure 9.2). This allowed them to make a compelling argument that a significant portion of Jasper's profits was directly attributable to the use of the stolen process. Figure 9.2 shows that customers indeed appreciated the quality of wine produced under VinLuxe IP, causing Kildare Wine's revenues to increase fro $400,000 annually to more than $750,000 annually.

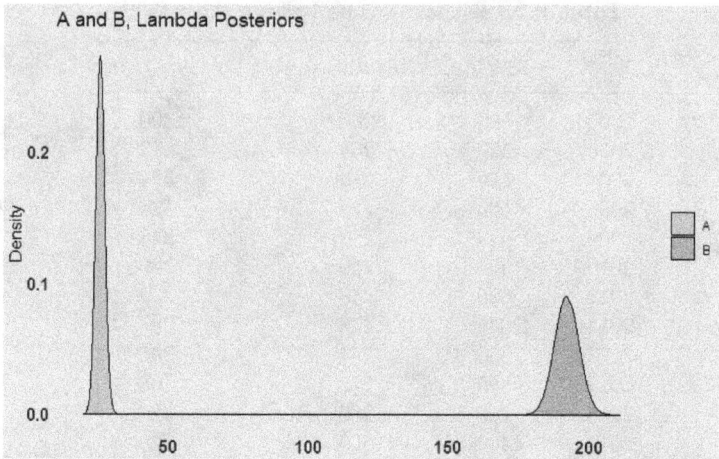

Figure 9.1. Kildare's net income (profits) before (A) and after (B) the alleged theft of trade secrets.

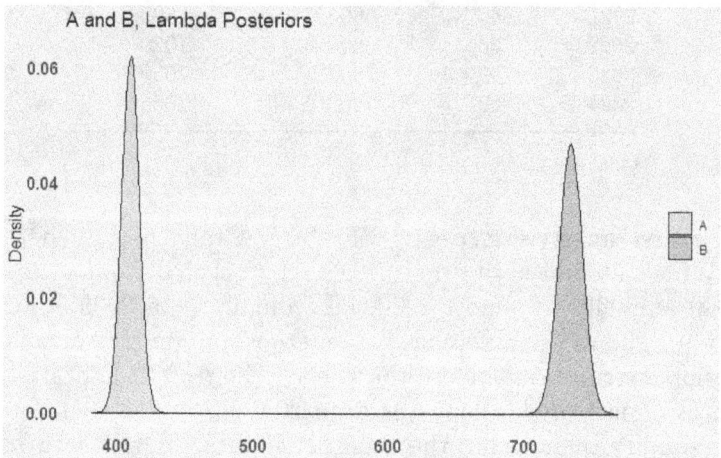

Figure 9.2. Kildare's revenue before (A) and after (B) the alleged theft of trade secrets.

Calculating the savings Jasper accrued from circumventing the research and development phase required a different approach. Elizabeth and her team analyzed the investment VinLuxe had made in developing their aging process, including costs related to labor, trials, errors, and time. They then used these figures to estimate the typical R&D expenditure required for developing a similar process.

The strategy was to demonstrate the cost disparity between Jasper's minimal investment in adaptation versus VinLuxe's substantial investment in creation. Expert witnesses from the industry testified to the extensive nature of such R&D efforts, helping to paint a vivid picture of the financial burden Jasper had unfairly avoided (Figure 9.3).

Jasper's operational savings were another critical area of focus. By adopting a process that had already been refined and optimized by VinLuxe, Jasper not only undercut the market but also significantly reduced his production costs. The VinLuxe team undertook a detailed analysis of Jasper's production expenses, comparing them with what was typical for a startup without access to established technology. Figure 9.3 shows that Kildare Wine's operating expenses dropped by around $30,000 after their acquisition of VinLuxe's IP, due to the substantial increase in efficiency, and reduction in waiting times to convert grape juice into fine wine.

To quantify these savings, they examined the efficiencies related to production volume, quality control, and speed to market, which Jasper had unfairly gained. The data collected provided a clear view of the cost differential, adding another layer to the financial claims against Jasper.

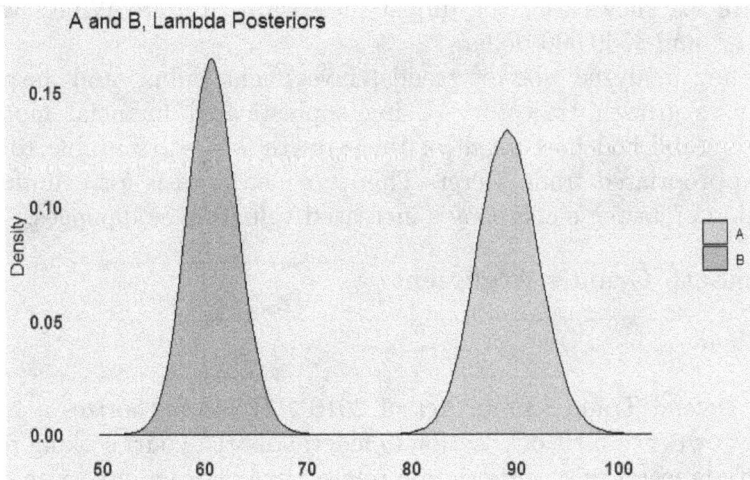

Figure 9.3.   Kildare's operating expenses before (A) and after (B) the alleged theft of trade secrets.

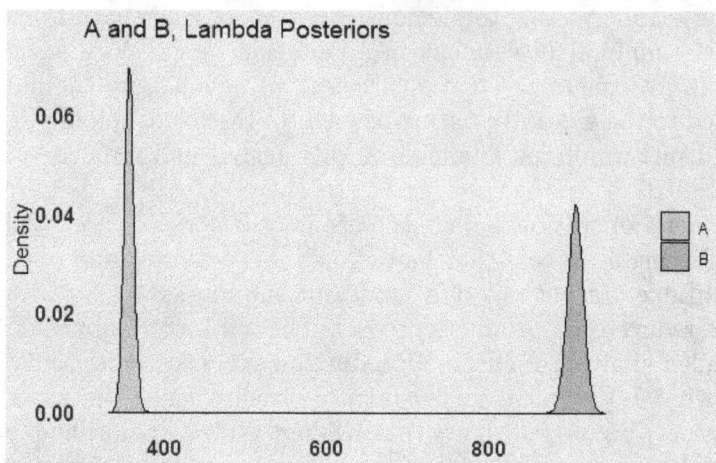

Figure 9.4. Kildare's net present value before (A) and after (B) the alleged theft of trade secrets.

Perhaps the most complex calculation was determining the increase in Jasper's business value resulting from the appropriated assets. Elizabeth's team collaborated with financial analysts and business valuation experts to assess the value of Jasper's company before and after the introduction of the stolen wine aging process. Figure 9.4 shows that the unjust increase in Kildare Wine's value was around $550,000 dollars

They analyzed market trends, investment influx, and Jasper's business growth trajectory. Using sophisticated financial models, they established how much of this growth was attributable to the misappropriated trade secret. This provided a basis for claiming a portion of Jasper's company's increased valuation as damages.

Elizabeth Grant's Argument

The Law

The Defend Trade Secrets Act of 2016 (DTSA) authorizes a trade secret owner to file a civil action in federal district court seeking relief for trade secret misappropriation related to a product or service used in or intended for use in interstate or foreign commerce (PL 114–153, 130 Stat 376 (2016)).

Allowable unjust enrichment damages under the DTSA are:

(1) the defendant's profits on sales resulting from the appropriated assets,
(2) the defendant's savings on R&D investments resulting from the appropriated assets,
(3) the defendant's savings from cost efficiencies resulting from the appropriated assets,
(4) the defendant's increased business value resulting from the appropriated assets.

Profits on sales: Are net sales revenues less costs that directly vary with sales (are directly attributable to sales). In the current assessment, Elizabeth Grant used 's reported net sales less their reported cost of goods sold.

R&D costs: Research and development benefits accrue to beneficiaries in subjective and sometimes indirect ways. This assessment chooses to render a conservative and completely objective assessment of Jasper Kildare's benefits from their IP theft, and avoids opining on the subjective yet potentially substantial benefits to Jasper Kildare from VinLuxe Innovations's customization, research and development.

Cost efficiencies: Were computed from the total of costs of goods sold plus selling, general and administrative costs.

Business value: was computed using the industry best-practice net present value (NPV) of free cash flows approach. Free cash flows were computed as net sales, less cost of goods sold, selling, general and administrative expenses and shareholder distributions. Jasper Kildare's cost of borrowed capital was used as the discount rate for the NPV.

Methodology

The following steps were taken to analyze the accounts and operations of Kildare Wine before and after the alleged theft of IP:

(1) Extraction of financial information from discovery documents, internal reports and external web scraping. This data was curated into a set of financial statements.

(2) The curated financial statements dataset contained periods and accounts with missing data. Elizabeth Grant used industry best-practice methods for both cross-sectional and time-series methods to interpolate missing data.

(3) All financial data was de-trended, to remove the impact of appreciating dollar amounts due to inflation and business growth. Without de-trending, business value and profits on sales after the alleged theft would be dramatically overstated, because the later numbers are substantially larger than earlier numbers. De-trending of data eliminates this problem and states all periods in the equivalent of 2015 (midpoint) dollars.

(4) Industry best practice Bayesian A/B tests were performed for the years before (B) versus the years after alleged theft of IP (A).

(5) The value added to Kildare Wine's business is computed for the relevant metrics from the posterior Bayesian means (lambda for the Poisson distribution) under scenario (A) versus scenario (B).

(6) The study maintained statistical control over "major factors" that could influence the dependent "unjust enrichment" variables by:

- consistently detrending financial figures, to control for inflation and Kildare Wine's organic growth, where the detrending used industry best-practice time-series models, and
- consistently testing for unobserved covariant predictors, through mixed-effects and other models where applicable.

(7) **Causation and Correlation:** the data analysis reveals strong correlation between Kildare Wine's misappropriation of IP and their subsequent unjust enrichment. Causation is also supported by the time sequence of events: after the alleged theft of Vin-LUse's IP, Kildare Wine benefited from increased profits on sales, substantial cost savings in operations, and increased enterprise value.

Specific interpretation and methodological choices in curating, analysis and calculation involving Kildare Wine's financial data has been motivated by:

- a choice of the most objective and least subjective methodology that will insulate data and analysis from investigator bias, while providing objective, verifiable and replicable conclusions,

- the elimination of effects due to confounding and unobserved variables, and
- the choice of the most appropriate methods and prior assumptions that would allow Kildare Wine's financial data, and the analysis based upon that data, to speak for itself without imposing any investigator bias.

Elizabeth Grant's Opinion

In my opinion, the sequence of tasks performed and methodologies used here are industry best-practice approaches to solving the specific analysis problems I encountered. The application of these methods has rendered conservative, objective and reliable estimates of unjust enrichment.

In my opinion, my calculations provide a reliable scientific basis, along with data supporting the opinion that each independent causal link presented in my analysis is true to a high degree of probability.

In my opinion, Kildare Wine unjustly enriched itself through the theft and use of VinLuxe Innovations intellectual property as follows:

(1) $330,000 of unjust enrichment from profits on sales resulting from the appropriated IP.
(2) $30,000 of unjust enrichment from reduced R&D investments as a result of appropriated IP.
(3) $175,000 of unjust enrichment from cost efficiencies which are reflected in excess net income that resulted from use of VinLuxe Innovations' appropriated IP.
(4) $550,000 of unjust enrichment from increased business value resulting from the appropriated IP.

Kildare Wine's $1,085,000 benefit from increased profits on sales, cost savings in operations, and increased enterprise value rationally flows from the theft of IP from VinLuxe Innovations and the other evidence presented. We seek restitution of damages in this amount.

Elizabeth Grant, PhD, CPA

Elizabeth Grant, PhD CPA

Armed with detailed financial analyses and backed by expert testimonies, Elizabeth presented a compelling case to the court. The evidence was laid out methodically, showing not just the direct losses but also the broader impacts of Jasper's actions on VinLuxe's business. Each claim was supported by robust data, turning abstract numbers into stark realities of financial damage.

As the trial progressed, the magnitude of Jasper's infringement became undeniable. The meticulous preparation by VinLuxe's legal team paid off as they painted a comprehensive picture of the financial ramifications of Jasper's theft.

The courtroom battles were punctuated with technical discussions on chemical compounds, molecular structures, and the arcane processes of wine aging. Expert witnesses from the fields of oenology, chemistry, and intellectual property law paraded before the judge, each adding their perspective to the layered narrative.

Outside the courtroom, the story captured the imagination of the public and the media. It was a tale of tradition versus technology, secrecy versus innovation. Debates raged in wine forums and among intellectual property experts about the implications of the case for the industry.

After months of legal skirmishes, the court ruled in favor of Vin-Luxe. It was determined that Jasper had indeed violated the DTSA by acquiring and using VinLuxe's trade secrets without consent. The court ordered an immediate cessation of Kildare Wine's operations and awarded substantial damages to VinLuxe for the breach of security and the potential risk posed to their business.

The outcome of the trial was a landmark victory for VinLuxe, setting a precedent in trade secret litigation. It was a testament to the importance of detailed financial forensics in proving the extent of damages and the efficacy of intellectual property laws. For VinLuxe, the victory was not just about monetary compensation but also about affirming the sanctity of innovation and the protection of ideas in the fiercely competitive wine industry. The verdict was a vindication for VinLuxe but a cautionary tale for the industry. It underscored the delicate balance between innovation and tradition, and the immense value — and vulnerability — of intellectual property in the modern age.

As for Jasper, he disappeared soon after the trial, leaving behind a legacy of scandal and a few thousand bottles of what was, by all accounts, spectacular wine. The story of his rise and fall was whispered in vineyards and savored like a fine wine, a blend of audacity and the age-old allure of a well-kept secret.

Chapter 10

Decisions with More than Two Choices

10.1. Bayesian Bandits

In A/B testing, researchers are typically presented with two data sets and two options: one for the treatment and one for the control. However, many real-world scenarios require more than two choices. In Chapter 5, the Locust Lane case study demonstrated the use of a multi-stage decision tree to address problems involving multiple options.

There is a more flexible and elegant solution for handling decision problems with three or more choices: multi-armed bandit approaches. These methods dynamically allocate resources to options that show superior performance while reducing allocations to those that underperform. This approach not only optimizes resource utilization but also accelerates the identification of effective choices, eliminating the need to wait for a clear winner to emerge.

To envision a bandit problem, consider the dilemma of a gambler in a casino faced with the option of choosing from N *one-armed bandit* slot machines. Each machine offers an undisclosed probability of winning. For instance, `Bandit 1` might have a winning probability of 0.8, while `Bandit 2` might have a winning probability of 0.4. The goal is to maximize winnings by consistently choosing the machine with the highest probability of winning. However, identifying which machine offers this highest probability presents a significant challenge.

Accurately determining the probability of a *winning* event from a particular machine requires a substantial number of data samples (i.e., pulls of the machine's arm); the greater the number of samples, the more precise the estimated probability becomes. Ideally, to achieve the highest level of accuracy, to reach maximum confidence in our estimates of each machine's probability of payout, we would need to take an infinite number of samples. That is the burden that frequentist statisticians must bear. Collecting a large number, such as one million samples, might seem sufficient for high confidence. Yet, if we were to engage one million times with less profitable machines in our search for the best bandit, we would end up playing inferior machines excessively. Thus, not only would we fail to maximize our winnings, but we might also incur significant losses.

Our gambler's predicament is known as the "explore vs. exploit" dilemma. Do we "explore" by trying new machines in hopes of discovering a higher winning probability? Or do we "exploit" by continuing with a machine that has already shown decent winning odds? This tension between the desire for accurate probability estimates and the goal to maximize returns from the highest-yielding bandit extends beyond casinos to many real-world applications.

For example, consider an online advertising platform aiming to display ads that users are most likely to click. Should the platform stick to ads that have already performed well, or should it test out new ads that could potentially yield even higher click-through rates? Similarly, news outlets like the *New York Times* face this issue when deciding whether to highlight popular articles that are known to attract readers or to introduce new content that could potentially engage more visitors.

Bayesians are realistic. They don't set themselves a goal of an infinite number of pulls on the bandit arms, and the certainty of financial ruin in the process, since the house always wins. Rather than pulling each lever thousands of times to ascertain exact probabilities, the Bayesian bandit strategy uses accumulated data to make more informed decisions about which machine to play. And it starts out with our informed or uninformed prior knowledge of the winning probabilities for each bandit. If a bandit seldom wins but has not been tried frequently, its winning probability is deemed low, yet our confidence in this estimate is also low. Therefore, it might still be worth giving this machine a few more tries. Conversely, if a machine is sampled extensively and often wins, then its high winning probability is deemed reliable, and it should be chosen more frequently.

This Bayesian approach optimizes our strategy by dynamically balancing between exploring new possibilities and exploiting known advantages, thus aiming to maximize overall returns in the face of uncertainty.

A fundamental aspect of bandit problems is that choosing an arm does not affect the properties of the arm or other arms. The solutions to such problems have many practical applications. They are useful where you are iteratively allocating a fixed, limited set of resources between competing choices in a way that minimizes the regret. It is a classic reinforcement learning problem of exploration versus exploitation. The crucial trade-off the gambler faces at each trial is between "exploitation" of the machine that has the highest expected payoff and "exploration" to get more information about the expected payoffs of the other machines [53].

Potential bandit applications are found in many fields, for example:

- **"Moneyball":** Which pitcher to invest in; you have annual contracts, so getting new information is slow and costly, and your net profit is the objective.
- **Clinical trials:** Is drug effective or not, where data is very costly, objective is profit (revenue - variable cost - R&D cost - discretionary cost).
- **Ad markets:** Targeting the most profitable ad market, and optimizing ad copy.
- **Police neighborhood deployment:** Decide where do you put your policing resources to minimize crime.
- **Portfolio choice:** In finance, each investible asset has an ROI distribution (unknown), and we want to know where to put our limited budget.

10.2. Case Study: Bandits at Research Arms ($A/B/C$ Decision Models)

Research Arms, Inc. is at the forefront of Earth's defense technology, playing a critical role in the escalating conflict over Uranus's vast resources. The planet, prized for its rich deposits of itanium — a mineral essential for Earth's advanced industrial applications — had become the center of a brutal war. The Klingons, from a distant planetary system, had set their sights on Uranus, driven by a fierce desire to control its itanium mines.

The stakes were high as Earth's Starfleet clashed with Klingon forces in the cold expanse of space. In response, Research Arms had developed three innovative weapons, each with the potential to decisively shift the balance in Starfleet's favor. The challenge now was to determine which weapon would most effectively neutralize the Klingon threat.

The weapon Number 1 was an ultraviolet laser, engineered to penetrate even the toughest Klingon shields. Its beams, tuned to a frequency specifically designed to disrupt Klingon technology, promised a lethal advantage in direct space combat.

The weapon Number 2 was a shaped charge nuclear device, capable of targeted destruction with unprecedented precision. This weapon was designed to annihilate key Klingon installations without causing extensive collateral damage to the itanium mining infrastructure.

The weapon Number 3 was the most unconventional: a psychological weapon. This device emitted a complex series of subsonic pulses and encoded messages that induced deep psychological distress and remorse in Klingons, exploiting their innate sense of honor and duty.

Preliminary studies and the opinions of Starfleet's experts suggested that weapon 1, the ultraviolet laser was effective only 20% of the time, due to Uranus' dense atmosphere. Weapon 2, the shaped charge nuclear device was 50% effective, being somewhat scattershot. Weapon 3, the psych weapon, was 70% effective on the highly conflicted Klingon personality. But these prior beliefs were only provisional, and more data was needed before a wholesale deployment of any one weapon.

Dr. Dina Moriv, the lead scientist at Research Arms, convened a team of experts to conduct a series of *simulations* and *Bayesian A/B/C testing* to evaluate the effectiveness of these weapons. The team included tactical analysts, weapons engineers, and a psychologist specialized in Klingon culture.

As the simulations commenced, the ultraviolet laser showed promising results by consistently breaching the simulated Klingon shields. However, its effectiveness was limited by the need for precise targeting and the considerable energy required to maintain its lethal output.

The shaped charge nuclear weapon proved devastatingly effective in the simulations, obliterating its targets with terrifying efficiency. However, there were concerns about the political and ethical implications of using such a weapon, as well as the potential for radioactive fallout that could render the itanium mines unusable.

The psychological weapon yielded the most uncertain results. Initial tests indicated that it could significantly demoralize Klingon forces, leading to retreats and surrenders in simulated scenarios. However, its effectiveness was highly variable, dependent on the psychological state and individual susceptibilities of the Klingon warriors.

Dr. Moriv and her team presented their findings in a high-stakes meeting with Starfleet Command. The room was tense, filled with top military officials and strategists, all aware that the decision they were about to make could end the war or lead to a prolonged, bloodier conflict.

Admiral Hayes, the commander of Starfleet operations, listened intently as Dr. Moriv outlined the capabilities and limitations of each weapon. The discussion was intense, with various factions advocating for different approaches based on strategic priorities and ethical considerations.

After hours of deliberation, a consensus began to emerge. While the ultraviolet laser and the shaped charge nuclear weapon had their merits, the psychological weapon held the potential not only to stop the Klingons but to do so in a way that aligned with Starfleet's values of minimizing loss of life and promoting peace.

Admiral Hayes made the final decision. "We will proceed with the deployment of the psychological weapon mixed with some use of the other two weapons. We will gather data on an ongoing basis through a specialized *Dashboard* to update our decisions over time. Prepare weapon 3, the psychological weapon, for immediate use and continue refining the other weapons as secondary options. Our goal is to end this conflict with as little bloodshed as possible, maintaining our moral integrity while securing Uranus's resources."

Research Arms Bayesian $A/B/C$ decision models were adapted from [54–56] and their decisions are presented in detail below (we have been able to replicate their classified R language code for Research Arms' $A/B/C$ decision models in the technical appendix at the end of this book).

10.2.1 *Success or failure of research arms weapon systems*

In the midst of the swirling chaos of war over the itanium mines on Uranus, Research Arms, Inc. stands as Earth's bulwark, crafting cutting-edge weaponry for the Starfleet. The company's strategy room buzzed with the electric intensity of a battlefield command center as they deployed their latest decision-making algorithm, a beacon of clarity in the fog of war.

This critical algorithm, adapted for the high-stakes environment of interstellar conflict, operates on principles akin to those found in the gambling parlors of old Earth, albeit vastly more sophisticated and deadly. Here's how the drama unfolds:

At each decision point, the algorithm sample a random variable X_b from the prior of bandit b, for all weapons choices b (i.e., where b is a single one out of Starfleet's three new experimental weapons). Each weapon, from lasers to psychological disruptors, is a bandit, vying for the chance to prove its mettle.

The system then selects the weapon with the highest potential impact, identified as the one with $maximum(X_b)$ value. This chosen 'bandit' is not merely a tool but a critical gambit in the intricate chess game against the Klingon adversaries.

With the weapon unleashed, the results are meticulously analyzed. The real-world performance data feeds back into the system, refining the probabilistic models that predict future success. This loop of action and learning is relentless, mirroring the ceaseless barrages of space warfare.

The cycle repeats, each iteration a pulse in the rapid heartbeat of conflict. The particular decision strategy used by Starfleet was first proposed by Canadian entomologist William Thompson in his 1937 book *Science and Common Sense: An Aristotelian Excursion*. Thompson's book was required reading for every cadet at Starfleet Acadamy, and Thompson Sampling was a fundamental skill understood by every burgeoning cadet. Thompson's approach was stripped of conventional complexities like Markov chain Monte Carlo methods or the esoteric minutiae of statistical fit measures, in order to cut through the Gordian knot of wartime decision-making with elegant simplicity.

Imagine, if you will, the practical application of this algorithm under the relentless stars. Three weapons systems, each with probabilities of turning the tide of battle — 20%, 50%, and 70%, respectively. Each "pull" of a weapon system is a gamble against fate, a roll of dice loaded with the gravity of solar empires. The reward mechanism is stark: the survival of Earth's outpost and the protection of its celestial mining assets.

The graphs in Figure 10.1 summarize what Starfleet found.

Consider how we would interpret these probability density graphs. The weapon (bandit) with the highest probability of winning, has a density that gets sharper and sharper as the number of trials increases. This varies quite a bit, and the densities are diffuse. *Weapon 3* is slightly sharper around 100 pulls of the arms, with

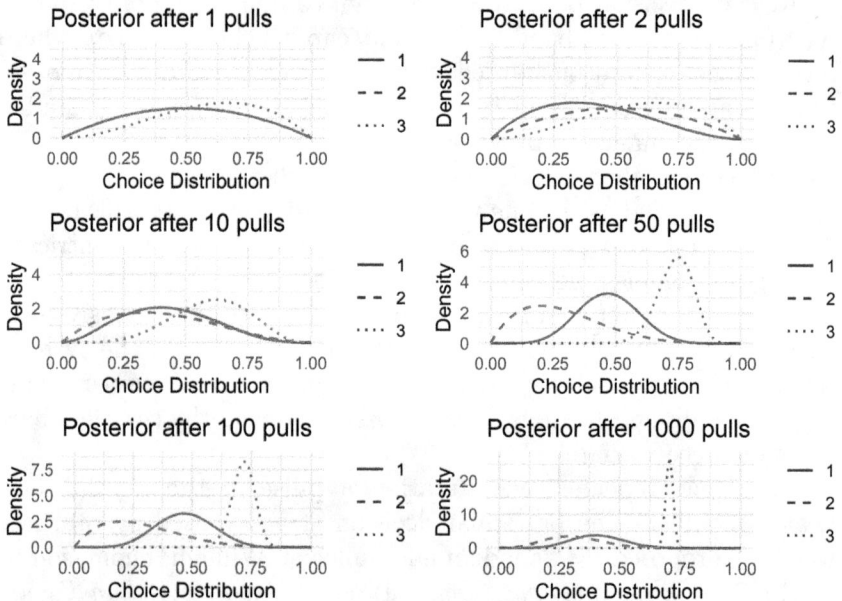

Figure 10.1. Choice probability distributions with 0.2, 0.5, and 0.7 probabilities of success for weapons (bandits) 1, 2, 3, respectively with increasing trials (pulls).

a reward of around 1.5. So once *Weapon 3* has "proved itself" we are much more likely to choose it in the future, and we just don't bother to choose the lower performing bandits, although there is a small chance we could.

A fat distribution means more "exploration". A sharp distribution means more "exploitation" (if it has a relative high win rate). Note that as the programmer, you don't choose whether or not to explore or exploit. You sample from the distributions, meaning it is randomly decided whether you should explore or exploit. Essentially, the decision is random, with a bias toward weapons that have proven themselves to win more.

10.2.2 *Weapons system effectiveness varies significantly with each use*

Amidst the raging space battle over the rich itanium mines of Uranus, Research Arms' weaponry faced an acute challenge. While each weapon in their arsenal displayed commendable average effectiveness, the results of their deployment told a more erratic tale — fluctuating success and unexpected failures against the Klingon encroachments.

The reality was stark; some weapons that seemed less potent on paper were occasionally yielding disproportionately high results, casting a shadow of unpredictability on strategic decisions. Such inconsistencies could no longer be overlooked, as the fate of the war hinged on every volley fired and every mine defended.

Recognizing the critical need for adaptability in strategy, the think tank at Research Arms undertook a comprehensive overhaul of their strategic algorithm. Their goal was to account not just for the average effectiveness of a weapon system but to embrace the inherent chaos of war — the randomness of outcomes that could turn an underdog weapon into a pivotal game-changer.

Research Arms' new algorithm began by factoring in the variability of weapon effectiveness as seen in previous engagements. This wasn't just about choosing the weapon with the best average outcome but also considering those with potential for high-impact results, even if less consistent. With this enhancement Starfleet's strategic command could deploy their arsenal not only based on historical averages but also informed by a calculated embrace of the unpredictable elements of warfare. The updated algorithm represented a bold leap into a form of combat analytics that could sway the course of the battle, leveraging both the science of probability and the art of war.

As the revised strategy rolled out, each deployment of Research Arms weaponry was a test of the new algorithm — each outcome a critical piece of the evolving war puzzle. Starfleet commanders watched anxiously as the new approach was put to the test, hoping that this blend of precision and adaptability would bring stability to the front lines and secure the valuable itanium resources crucial for Earth's future.

Each weapon system offered unique capabilities and outcomes in the theater of war. The psychological weapon, known as Psyche Pulse One, had initially shown promise. This sophisticated device aimed to exploit the Klingon's inherent sense of honor, inducing deep remorse for their aggressive actions. Although effective, its impact was gradual, and the warlike Klingons, with their resilient psyche, often recovered swiftly from its effects, resuming their assaults with even greater ferocity.

In contrast, the second weapon in the arsenal, the Nuclear Disruptor, wielded a more immediate and *twice* as potent force compared to Psyche Pulse One. It was designed to deliver precise, high-yield blasts capable of obliterating Klingon outposts and armadas in a single, thunderous roar. Its effectiveness was undeniable, making it a critical asset in Starfleet's tactical operations.

Yet, it was the third weapon that became the legend whispered across starship corridors and feared by all who faced it. The "Laser", once fully operational and piercing through the dense, toxic clouds of Uranus, proved to be a game-changer. It was *three* times more effective than the psychological weapon and had a devastating effect on the Klingon forces. Wherever the laser struck, nothing remained of the vaporized Klingons.

As the war continued, the strategic decisions became more complex. The balance of power shifted with each encounter, each skirmish in the shadow of Uranus. The stakes were astronomical — a planet's worth of resources and the strategic upper hand in an interstellar conflict.

Figure 10.2 shows the results of the revised algorithm that Starfleet used to incorporate the different effectiveness (i.e., payout) that existed on each trial (pull) of each of the three weapon systems.

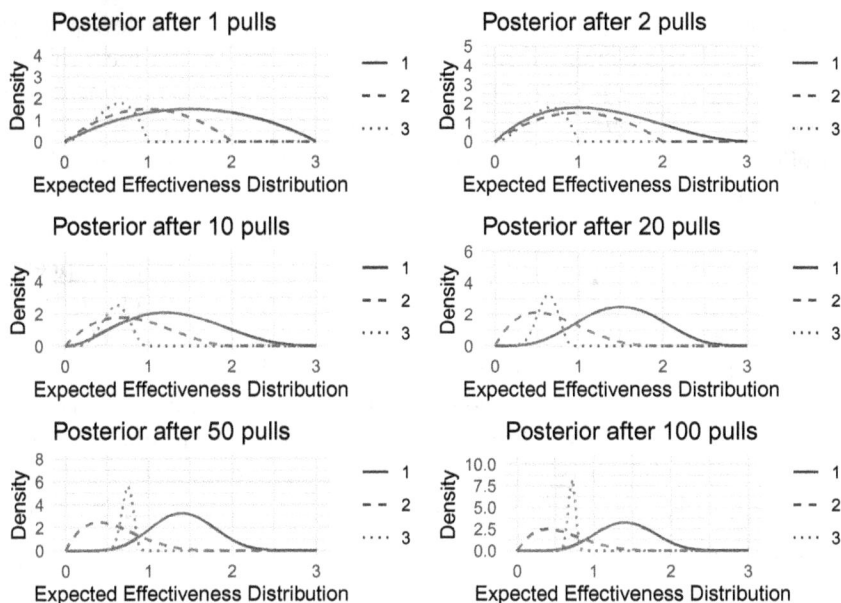

Figure 10.2. Choice probability distributions with 0.2, 0.5, and 0.7 probabilities of success with effectiveness (payout) of 3, 2, and 1, respectively.

As the laser systems were deployed into the orbit of Uranus, hidden beneath layers of protective shields, the anticipation was palpable. These engineers had argued, amidst much skepticism, that the raw power of the lasers — even with their sporadic accuracy — would prove far more crippling to the Klingon fleets than the more reliably successful psychological weapons. These psych weapons, designed to exploit the Klingon's emotional and mental states, had indeed shown a high success rate, manipulating their feelings of remorse and honor to great effect. However, their impact, while significant, lacked the sheer annihilative force that the engineers believed was necessary to decisively end confrontations.

The moment of truth came during a fierce skirmish near the Cytherean Gap, a strategic passageway close to the itanium mines. Starfleet's fleet, under heavy fire, unleashed the full fury of their laser systems. The beams, powered by the latest advancements in photon acceleration, cut through the toxic clouds of Uranus with unerring precision that belied their previous inconsistencies (Figure 10.2).

In mere moments, the space was illuminated with the devastating brilliance of the lasers. Klingon warships, caught in the path of the beams, were vaporized into stardust, their debris scattering across the void. The impact was undeniable — each successful strike multiplied in effect, creating a spectacle of destruction that turned the tide of the battle.

Back at Research Arms, Inc., the engineers watched the live feeds, their faces a mix of awe and relief. Once questioned by their peers, they were indeed proven prescient. The lasers, though few in their hits, delivered catastrophic blows to the Klingon forces, dwarfing the psychological warfare's impact in sheer visceral effect.

This pivotal victory marked a new chapter in the conflict over Uranus. Research Arms, Inc. continued to refine and enhance their laser systems, driven by the vindication of their engineers' insights.

10.2.3   *Count data*

In the deep, shadowy expanse of space surrounding the tumultuous skies of Uranus, the war for control over the invaluable itanium mines had escalated to unprecedented levels. Amidst the chaos, an urgent directive was issued by Starfleet Command, echoing through the corridors of Research Arms, Inc.'s heavily fortified orbital command center.

As laser beams sliced through the toxic clouds and explosions cast eerie lights on the battlefield, a new mandate emerged — count the Klingon casualties. This grim task wasn't just a morbid necessity; it had become a crucial metric, a "figure of merit", in Starfleet's intensifying public relations effort to justify the war back on Earth. The conflict was draining, sapping not only vast financial resources but also claiming the lives of countless brave souls.

Starfleet Command, determined to maintain public support, aimed to demonstrate their effectiveness in winning not just the strategic locations but also the "hearts and minds" of the local Uranian population. They portrayed their efforts as a crusade to liberate the region from Klingon tyranny, emphasizing the high number of Klingon combatants eliminated as a measure of their success.

As the counts began, the initial tallies were relayed back to Earth, feeding a voracious media landscape eager for updates. However, the complexity of accurately assessing such figures under the fog of war led Research Arms' analytics division to a crucial realization: their current models were inadequate for the task at hand. The situation demanded more than just rudimentary counting; it required sophisticated predictive modeling to ensure accuracy and reliability in the chaotic war environment and a choice of weaponry through Bayesian $A/B/C$ decision modeling.

The results of this exhaustive analysis were meticulously compiled and presented in Figure 10.3, where weapon systems 1, 2 and 3 were shown that average 1, 2 and 3 Klingon kills per minute.

This addresses the expected rewards as derived from counts based on the Poisson distribution. It is important to contextualize these findings and acknowledge the inherent instability in determining winners in such scenarios. Unlike a straightforward [0,1] outcome distribution or its Beta prior, our analysis employs a more complex Gamma prior for the counts, which allows for a range from [0, max]. This variability is a primary reason for the observed instability in outcomes.

Interpreting the Gamma density can be challenging, suggesting a potential advantage in using simpler Bernoulli trials for this type

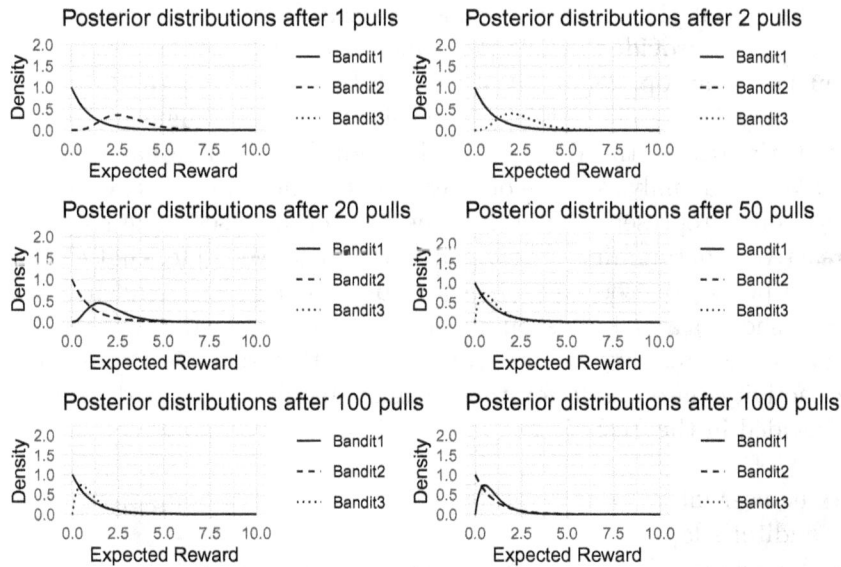

Figure 10.3. Choice probability distributions where wapons 1, 2 and 3 had average 1, 2 and 3 Klingon kills per minute.

of analysis. Bernoulli trials provide a clearer binary outcome, which might be more stable and interpretable for decision-making purposes.

It should be noted that the company initially experimented with count data but ultimately found it too unstable for reliable corporate decision-making. This led to the decision to reject this approach in favor of methodologies that ensure greater consistency and predictability in the outcomes, essential for strategic applications.

10.3.   Secure Databases using a Real-time Dashboard

Weapons designers, auditors, lawyers, social researchers and, indeed, even the war-mongering Klingons face a particular problem in that much of the information they need may be subject to intense security, in proprietary client files, in central locations, behind firewalls, and on powerful servers or cloud platforms. To secure sensitive or proprietary information and address privacy concerns, it is important that only the smallest amount of necessary information be maintained on mobile platforms such as laptops that are used in the field.

The standard solution to such problems is found in client–server systems that place high-performance, secure systems on a centralized server, and provide the field personnel with light, client software that runs on a laptop, communicating with the server over the Internet.

Shiny is the client–server extension of the R language, and as with the rest of the R language, is uniquely suited to handling the *ad hoc* data analysis needs of researchers in the field, where each new task often represents an entirely new set of analyses. It implements reactive programming, which starts with reactive values that change in response to user input (such as positioning the confidence and cost sliders) and builds on top of them with reactive expressions that access reactive values and execute other reactive expressions. Reactivity based code for the Bayesian bandit dashboard below is provided in the Technical Appendix. This has two parts:

- the user interface `ui` which conceivably would operate on the field auditor's laptop, and
- the server-side `server` operations that would take place at the audit firms headquarters behind a firewall, and with access to firm and client files.

Shiny is also a tool for fast prototyping of digital dashboards, giving you a large number of HTMLWidgets at your disposal which lend themselves well to building general-purpose web applications. Shiny is particularly suited for fast prototyping and is fairly easy to use for someone who is not a programmer. Dashboards locally display some data (such as in a database or a file) providing a variety of metrics in an interactive way.

Figure 10.4 only shows the user interface dashboard created by the ui side of this application. For the full application, refer to the Technical Appendix at the end of this book. Note that to run a Shiny app, it needs to be run as a script (this is the way that the code in the appendix is set up) or it needs to be split into the ui and server portions to be run in a client-server arrangement. POSIT, the main organization supporting R, offers a free client-server platform at https://www.shinyapps.io/ which allows small demonstration apps to be run, and is appropriate for the Shiny code example in this book.

Research Arms' engineers implemented the previous bandit models in a Shiny app so that as new data was received, for example with every firing of the laser, or following the receipt of new counts of Klingons vaporized, they could dynamically update their posteriors

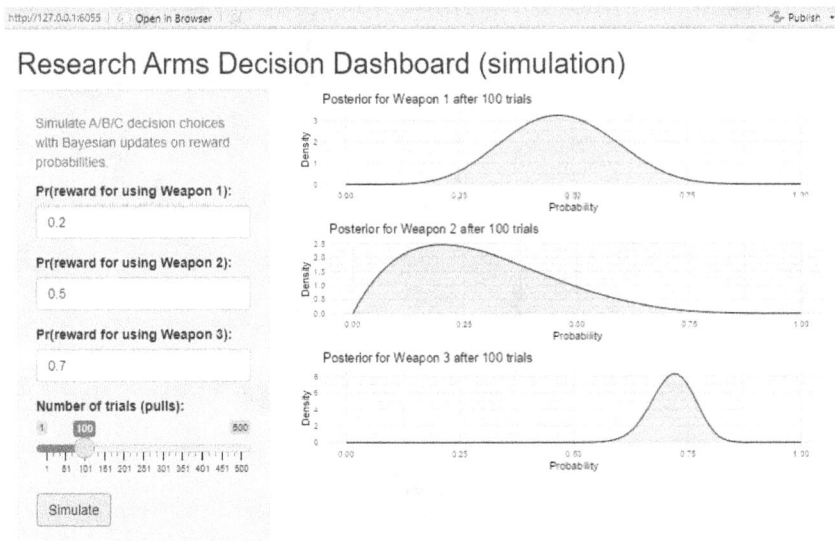

Figure 10.4. Research Arms' Bayesian multi-arm Bandit Simulation.

in real-time. Figure 10.4 shows the three arms representing decisions on three different weapon systems, showing a simple interface that allows simulated updating of priors and trials. This time the Psyche Pulse One had clearly outperformed the other two weapons.

Dr. Moriv felt a mix of relief and apprehension as she returned to her lab to oversee the final adjustments to the psychological weapon. The success of her invention was now tied to the fate of a planet, the future of its people, and the hope of peace in a war-torn galaxy.

As Starfleet deployed the weapon against the Klingon forces orbiting Uranus, the initial reports were cautiously optimistic. Klingon commanders were indeed retreating, confused and demoralized by the unexpected assault on their psyche.

The war for Uranus was far from over, but thanks to the ingenuity and dedication of Research Arms, Inc., Earth's Starfleet had a new weapon in its arsenal — one that wielded the power of remorse to quell the fires of war.

Chapter 11

Final Thoughs

Over a century ago, scholars like Ronald Fisher, Karl Pearson, J.B.S. Haldane, and William Sealy Gosset devised the frequentist statistical decision framework used up to the current day. Working with nothing more than pen, paper, and rudimentary adding machines, these pioneers crafted their mathematics to be as simple as possible, often at the expense of accuracy. They tackled the complexities of their era with squared-error loss functions and optimization through differential calculus, a testament to their ingenuity and dedication.

In their time, these brilliant minds were not just solving statistical problems; they were framing a new perspective on the world, one calculation at a time. Their objectives might have been different, reflecting the unique challenges and limited technological resources of their age, but their legacy is undeniable.

Fast forward to today, and the landscape of statistical modeling has transformed dramatically. Armed with virtually unlimited computational power and expansive memory capacities, we stand on the shoulders of these giants, exploring a vast array of possibilities that extend far beyond their imaginations and methods. Yet too often, out of habit or tradition, we cling to antiquated and inefficient methods.

In particular, the cornerstone of modern statistical A/B decisions is the Neyman–Pearson Lemma, foundational to the frequentist approach that dominates contemporary A/B testing scenarios. Statistician Bruce Hill, who was one of my PhD mentors, would illustrate the inherent limitations of such frequentist approaches with examples like the Behrens-Fisher problem, which is deemed insurmountable from a frequentist perspective, but is not a problem at all when viewed through a Bayesian lens. Bayesians embraces a powerful, flexible approach to statistics that asks "How much have we learned?" rather than "What is the absolute truth?" Absolute truths are elusive, and prejudiced by the questions we ask, the things we measure, and the limitations of our finite minds. But Bayesians know that there is always the opportunity to learn. Bayesians view every new piece of data as part of a never-ending puzzle, a chance to learn and refine our understanding incrementally.

Our exploration into bandit problems revealed that while traditional frequentist methods might reach approximate solutions if one has sufficient computational firepower, Bayesian methods offer an intuitive, straightforward way to make decisions. Bayesians appreciate the inherent uncertainties of the world, recognizing that not everything can be known. Yet, with controlled data collection at manageable costs, Bayesians continually enhance their understanding, tuning the pace and quality of their learning to meet practical needs.

Our journey into the science of informed decision-making has been as much about the methods we use to summarize and interpret data as it has been about the philosophical underpinnings that guide these choices. Bayesian innovations in A/B decision models not only allow us to utilize all available data but also enable us to learn efficiently from this data and make the best possible choices.

As we close this book, remember that this is merely the beginning of your own statistical voyage. Armed with the examples and ideas from this book, you are now equipped to navigate the complexities of informed decision in an environment that embraces continual learning. You are the new generation that will carry forward the spirit of curiosity and the legacy of learning that has been handed down through generations of statisticians.

## Chapter 12

Technical Appendix

12.1. R Code for Chapter 1

There is no code for chapter 1.

12.2. R Code for Chapter 2

```r
# Load the required packages
library(bayesAB)
library(dplyr)
library(ggplot2)

# Set seed for reproducibility
set.seed(123)

# Simulate data for group A
n_A <- 1000
conversion_rate_A <- 0.40
data_A <- rbinom(n_A, 1, conversion_rate_A)

# Define the range of conversion rates for group B
conversion_rate_B_values <- seq(0, 1, by = 0.01)
n_B <- 1000
```

```
# Initialize vectors to store results
p_values <- A_est <- B_est<-
  numeric(length(conversion_rate_B_values))
bayesian_probabilities <-
  numeric(length(conversion_rate_B_values))

# Informative priors
prior_alpha_A <- 10   # Prior successes
prior_beta_A <- 90    # Prior failures


# Loop over each conversion rate for group B
for (i in seq_along(conversion_rate_B_values)) {
  conversion_rate_B <- conversion_rate_B_values[i]

  # Generate data for group B
  data_B <- rbinom(n_B, 1, conversion_rate_B)

  # Frequentist £A/B£ test
  freq_test <- prop.test(c(sum(data_A),
                           sum(data_B)),
                         c(n_A, n_B))
  p_values[i] <- freq_test$p.value
  A_est[i] <- freq_test$estimate[[1]]
  B_est[i] <- freq_test$estimate[[2]]

  # Bayesian £A/B£ test
  bayes_test <- bayesAB::bayesTest(
    data_A,
    data_B,
    priors = c('alpha' = prior_alpha_A,
               'beta' = prior_beta_A),
    distribution = 'bernoulli'
  )

  # Extract the probability that B is better than A
  bayesian_probabilities[i] <-
    max(bayes_test$posteriors$Probability$B)
```

```
}

# Create a data frame for plotting
results_df <- data.frame(
  conversion_rate_B =
    conversion_rate_B_values,
  p_value = p_values,
  bayesian_probability =
    bayesian_probabilities,
  A_values = A_est,
  B_values = B_est
)


ggplot(results_df) +
  geom_line(aes(x = conversion_rate_B,
                y = p_value), color = 'darkgrey') +
  geom_line(aes(x = conversion_rate_B,
                y = bayesian_probability),
            color = 'black') +
  geom_hline(yintercept = 0.4,
             linetype = 'dashed',
             color = 'lightgrey') +
  geom_vline(xintercept = 0.4,
             linetype = 'dashed',
             color = 'lightgrey') +
  labs(title = " ",
#'Bayesian £A/B£ Probabilities (black) versus
#'Frequentist p-values (grey)',
       x = 'Conversion Rate B',
       y = 'Probability B > A & p-value') +
       theme_bw()

library(ggplot2)
library(tidyr)
library(reshape2)
library(tidyverse)
library(kableExtra)
```

```r
stats <- data.frame()
hypt <- vector()
for (j in 1:2000) {
  hypt[j] <- i <- 0
  for (i in 1:100) {
    k <- j * 5
    A <- rnorm(k, 110000, 300000)
    B <- rnorm(k, 100000, 300000)
    if (t.test(x = A, y = B,
               paired = TRUE)$statistic < 0)
      hypt[j] <- hypt[j] + 1
    stats[i, j] <- t.test(x = A, y = B,
                          paired = TRUE)$p.value
  }
}

meanp <- apply(stats, 2, mean)
mp <- data.frame(meanp, seq(5, 10000, 5))
colnames(mp) <- c("p", "sample_size")

hyp <- cbind(hypt, seq(5, 10000, 5)) %>%
  as.data.frame()
colnames(hyp) <- c("pct_false", "sample_size")
to_graph <-
  inner_join(mp, hyp,
             by="sample_size") %>%
  mutate(pct_false = pct_false/100)

long_graph <- pivot_longer(to_graph,
                           cols = c("p","pct_false"))
long_graph <- long_graph %>%
  mutate(name = ifelse(name == "p",
                       "p-value",
                       "Wrong Decisions"))
ggplot(data=long_graph, aes(x=sample_size,
                            y=value,
                            colour=name)) +
  geom_smooth() +
```

```
    scale_colour_grey(start = 0, end = .6) +
  theme_bw() +
# ggtitle("More Information Lowers the Proportion of
           Wrong Decisions") +
  xlab("Sample Size") +
  ylab("Proportion") +
  theme(legend.title=element_blank())
```

12.3. R Code for Chapter 3

```
library(bayesAB)
library(tidyverse)

# Set the lambda values for datasets A and B
lambda_A <- 45
lambda_B <- 46

# Define the size of each dataset
size_A <- 1000
# Number of data points in dataset A
size_B <- 1000
# Number of data points in dataset B

# Generate Dataset A with Poisson distribution
dataset_A <- rpois(size_A, lambda_A)

# Generate Dataset B with Poisson distribution
dataset_B <- rpois(size_B, lambda_B)

# Optionally shuffle data to randomize the order
set.seed(42)  # Setting seed for reproducibility
dataset_A <- sample(dataset_A)
dataset_B <- sample(dataset_B)
```

```
# Scenario 1: Weak prior
test_weak_prior <- bayesTest(
  A = dataset_A,
  B = dataset_B,
  priors = c("shape" = 1, "rate" = 1),
  dist = 'poisson'
)

plot(test_weak_prior)
```

12.4. R Code for Chapter 4

```
library(tidyverse)
library(ggplot2)

# Define the parameters for the two normal distributions
mean1 <- 600
sd1 <- 200
mean2 <- 400
sd2 <- 200


q_B <- qnorm(seq(0,1,.01),
             mean = mean2, sd = sd2)
sq <- seq(0,1,.01)
q_B <- cbind(sq,q_B) %>%
  as.data.frame()

# prob=.85 implies .15 in tail


q_A <- qnorm(seq(0,1,.01),
             mean = mean1, sd = sd1)
q_A <- cbind(sq,q_A) %>%
  as.data.frame()
```

```r
## prob=.98 implies .02 in tail


# Create a sequence of x values
x <- seq(00, 1500, length.out = 1000)

# Calculate the density values for both distributions
density1 <- dnorm(x,
                  mean = mean1,
                  sd = sd1)
density2 <- dnorm(x,
                  mean = mean2,
                  sd = sd2)

# Create a data frame to hold the values
df <- data.frame(x = c(x, x),
                 density =
                   c(density1, density2),
                 group =
                   factor(rep(c("Choice A",
                               "Choice B"),
                             each = length(x))))

# Plot the density functions
ggplot(df, aes(x = x, y = density, color = group)) +
  geom_line(size = 1) +
  scale_color_manual(values = c("grey", "black")) +
  theme_minimal() +
  geom_vline(xintercept = 800,
             linetype = "dashed") +
  labs(title =
         "Density Functions of Two Posterior
         Distributions",
       x = "Unit sales",
       y = "Density") +
  theme(legend.title = element_blank()) + xlim(0,1500)
```

```r
library(tidyverse)
library(ggplot2)

# Define the profit function
profit_function <- function(units) {
  profit <- 4 * units - 1000
  return(profit)
}

# Define the density functions for the two distributions
density_1 <- function(x) {
  dnorm(x, mean = 600, sd = 200)
}

density_2 <- function(x) {
  dnorm(x, mean = 400, sd = 200)
}

# Generate data for plotting
units_sold <- seq(0, 1000, by = 1)
profit_1 <-
  density_1(units_sold) *
  profit_function(units_sold) * units_sold
profit_2 <-
  density_2(units_sold) *
  profit_function(units_sold) * units_sold

# Create a data frame for ggplot
data <- data.frame(
  Units = rep(units_sold, 2),
  Profit = c(profit_1, profit_2),
  Choice = rep(c("Choice A", "Choice B"),
               each = length(units_sold))
)

# Plot the data
ggplot(data,aes(x = Units,y = Profit,
       color = Choice)) +
```

```
geom_line(lwd=1) +
scale_color_manual(values = c("grey50", "grey80")) +
labs(
  title = "Expected Total Profit for
           Number of Units Sold",
  x = "Number of Units Sold",
  y = "Expected Total Profit",
  caption = "Profit maximizing choice"
) +
theme_minimal()
```

12.5. R Code for Chapter 5

```
library(tidyverse)
library(bayesAB)

set.seed(123456)

A_MW_IMDB <- rpois(10000, 2)
B_MotB_IMDB <- rpois(10000, 3)

AB_IMDB <- bayesTest(A_MW_IMDB, B_MotB_IMDB, priors =
                     c('shape' = 4, 'rate' = 4),
                     distribution = 'poisson')

A_MW_RT <- rpois(10000, 3)
B_MotB_RT <- rpois(10000, 2)

AB_RT <- bayesTest(
  A_MW_RT,
  B_MotB_RT,
  priors = c('shape' = 4, 'rate' = 4),
  distribution = 'poisson')
```

```
AB <- combine(AB_IMDB, AB_RT, f = `+`)

plot(AB)
```

12.6.   R Code for Chapter 6

```
library(bayesAB)
library(tidyverse)
library(readr)

coin <-
  read_csv("DATABASE_xisBTC_yisETH_coin.csv") %>%
  select(btc=High.x, eth=High.y, date=Date)

# Example price series for Asset A and Asset B
price_btc <-  coin$btc
#  c(100, 102, 105, 110, 108, 115)
                                # Example prices for Asset A
price_eth <- coin$eth
# c(50, 52, 55, 60, 58, 65)
                                # Example prices for Asset B

compute_incremental_roi <- function(prices) {
  # Calculate ROI between each successive data point
  roi <- diff(prices) / head(prices, -1)
  return(roi)
}

# Compute ROI for both series
roi_a <- compute_incremental_roi(price_btc)
roi_b <- compute_incremental_roi(price_eth)


df_btc <- data.frame(roi_a,"BTC")
colnames(df_btc) <- c("ROI", "Cryptocurrency")
df_eth <- data.frame(roi_b,"ETH")
```

```r
colnames(df_eth) <- c("ROI", "Cryptocurrency")
df_coin <- rbind(df_btc,df_eth)

df_coin %>% ggplot(aes(ROI,color=Cryptocurrency))+
  geom_density(lwd=1)+theme_minimal()

# Perform the Bayes £A/B£ test
ab_test_result <- bayesTest(
  A = roi_a,
  B = roi_b,
  priors = c("mu" = 0,
             "lambda" = 1,
             "alpha" = 2,
             "beta" = .05),
  n_samples = 1e+05,
  dist = 'normal'
)

plot(ab_test_result)
#summary(ab_test_result)


library(tidyverse)
library(reshape2)
library(readr)


portfolio <-
  read_csv("DATABASE_portfolio.csv")[-60,]
# obs 60 is an average
portfolio[is.na(portfolio)] <- 0

compute_incremental_roi <- function(prices) {
  # Calculate ROI between each successive data point
  roi <- diff(prices) / head(prices, -1)
  return(roi)
}

roi <-
```

```r
  apply(portfolio[,2:10], 2, compute_incremental_roi) %>%
  as.data.frame() %>%
  cbind(portfolio$Year[2:59]) %>%
  .[-1,]

# Define shades of grey and linetypes
grey_shades <- c("black",
                 "grey80", "grey50",
                 "grey20","grey60","grey30")
line_types <- c("solid", "longdash",
                "dotted", "twodash",
                "dotdash", "F1")

melt(roi[,1:6,10]) %>%
  ggplot(aes(value, color=variable,
             linetype = variable)) +
  geom_density(size=.5) +
  scale_color_manual(values = grey_shades) +
  scale_linetype_manual(values = line_types) +
  theme_minimal() + xlim(-.5,.5) +
   theme(legend.position = "top",
         text = element_text(size = 12),
         panel.grid.major = element_blank(),
         panel.grid.minor = element_blank()) +
   labs(title = "Densities of Asset Price ROIs",
        x = "Value", y = "Density")


library(tidyverse)
library(kableExtra)
library(bayesAB)


library(fitdistrplus)
fitto <- tibble()
for(i in 1:6){
  fit <- tibble(
    colnames(roi)[i],
    descdist(roi[,i],
```

```
            method='unbiased',
            print=, graph=F)$mean,
    descdist(roi[,i],
            method='unbiased',
            print=F, graph=F)$sd,
    descdist(roi[,i],
            method='unbiased',
            print=F, graph=F)$skewness,
    descdist(roi[,i],
            method='unbiased',
            print=F, graph=F)$kurtosis
  )

  fitto <- rbind(fit, fitto)
}

colnames(fitto) <- c("Asset",
                    "Mean", "SD",
                    "Skewness", "Kurtosis")

fitto %>% kable(
  caption = "Fit statistics of
  each asset's time series",
  digits=3,
  ,booktabs=T
  )




table <- tibble()
   colnames(table) <- c("assetA",
                        "assetB",
                        "meanA",
                        "meanB")
for (j in 1:5) {
  for (k in (j + 1):6) {
    roi_test_result <- bayesTest(
```

```
    A = roi[, j],
    B = roi[, k],
    priors = c(
      "mu" = 0,
      "lambda" = 1,
      "alpha" = 2,
      "beta" = .05
    ),
    n_samples = 1e+05,
    dist = 'normal'
)

line <- data.frame(colnames(roi)[j],
                   colnames(roi)[k],
                   mean(t(as.numeric(
  unlist(roi_test_result$posteriors$Mu$A)
))), mean(t(as.numeric(
  unlist(roi_test_result$posteriors$Mu$B)
)))))
colnames(line) <- c("assetA",
                    "assetB",
                    "meanA",
                    "meanB")
  table <- rbind(table, line)

  }
}

  result <- max(unlist(table[,3:4]))
  winner <- table %>%
    filter(meanA == result) %>% .[1,c(1,3)]
  winner %>% kable(
    caption="The asset class with the highest ROI",
    digits=5,
    col.names =
      c("Asset", "ROI"),
    booktabs=T
      )
```

12.7. R Code for Chapter 7

```
library(tidyverse)
library(bayesAB)

set.seed(123)
A_binom <- rbinom(100, 1, .45)
B_binom <- rbinom(100, 1, .49)

AB1 <- bayesTest(A_binom,
                 B_binom,
                 priors =
                     c('alpha' = 1, 'beta' = 1),
                 distribution = 'bernoulli')

# summary(AB1)
plot(AB1)


set.seed(123)
A_norm <- rnorm(100, 6, 1)
B_norm <- rnorm(100, 5, 5)


AB2 <- bayesTest(
  A_norm,
  B_norm,
  priors = c(
    'mu' = 5,
    'lambda' = 1,
    'alpha' = 3,
    'beta' = 1
  ),
  distribution = 'normal'
)
```

```
AB3 <-
  combine(
    AB1,
    AB2,
    f = `*`,
    params = c('Probability', 'Mu'),
    newName = 'Expectation'
  )
##print(AB3)
#summary(AB3)
plot(AB3)
```

```
set.seed(456)
A_binom <- rbinom(100, 1, .35)
B_binom <- rbinom(100, 1, .30)
A_norm <- rnorm(100, 5.5, 4)
B_norm <- rnorm(100, 5, 3)


## Summarize the posterior means
#and variances from the first experiment,
#and insert these into the priors of the new £A/B£ test

mu1 <- mean(c(AB1$posteriors$Probability$A,
             AB1$posteriors$Probability$A))
var1 <- var(c(AB1$posteriors$Probability$A,
             AB1$posteriors$Probability$A))

## From Table 2 in Chapter 3 on Priors

alpha1 <-  (mu1^2-mu1^3-mu1*var1)/var1
beta1 <-  (mu1-1)*(mu1^2-mu1+var1)/var1


mu2 <- mean(c(AB2$posteriors$Sig_Sq$A,
             AB2$posteriors$Sig_Sq$B))
```

```
var2 <- mean(c(AB2$posteriors$Sig_Sq$A,
               AB2$posteriors$Sig_Sq$B))


## From Table 2 in Chapter 3 on Priors
lambda2 <- 1; alpha2 <- 2
mu2 <- mu2
beta2 <- var2


AB1a <- bayesTest(A_binom,
                  B_binom,
                  priors =
                    c('alpha' = alpha1,
                      'beta' = beta1),
                  distribution = 'bernoulli')

AB2a <- bayesTest(
  A_norm,
  B_norm,
  priors = c(
    'mu' = mu2,
    'lambda' = lambda2,
    'alpha' = alpha2,
    'beta' = beta2
  ),
  distribution = 'normal'
)

AB3 <-
  combine(
    AB1a,
    AB2a,
    f = `*`,
    params = c('Probability', 'Mu'),
    newName = 'Expectation'
  )
#print(AB3)
```

```r
#summary(AB3)
plot(AB3)


library(tidyverse)
library(EnvStats)  # for the Pareto distribution


set.seed(123)
dt <- sample(seq(as.Date('2024-02-01'),
                 as.Date('2024-05-31'), by = "day"),
                 10000, replace = TRUE)

NR <- rbernoulli(10000, .3) %>%
  as.data.frame()
colnames(NR) <- "type"
NR <- NR %>%
  mutate(
    userType = ifelse(type == TRUE,
                      "New Visitor",
                      "Returning Visitor"))

NR <- cbind(NR,dt)
NR_T <- NR %>% filter(type == TRUE)
NR_F <- NR %>% filter(type == FALSE)

## Don't differentiate between the userType;
# let set.seed() preselect whether A or B is the winner
sales_T <- abs(rnorm(nrow(NR_T),
                     25,100)) %>%
  floor()
sales_F <- abs(rnorm(nrow(NR_F),
                     25,100)) %>%
  floor()

daysSinceLastSession_T <- 0
daysSinceLastSession_F <-
  rpareto(nrow(NR_T), 1, .5)
daysSinceLastSession_F <-
```

```
  ifelse(
    daysSinceLastSession_F > 100,
    runif(1, 1, 5),
    daysSinceLastSession_F) %>% floor()

nrt <-   cbind(NR_T,
                cbind(sales_T,
                        daysSinceLastSession_T)) %>%
  rename(
    sales = sales_T,
    daysSinceLastSession =
      daysSinceLastSession_T)

nrf <-   cbind(NR_F, cbind(sales_F,
                        daysSinceLastSession_F)) %>%
  rename(
    sales = sales_F,
    daysSinceLastSession =
      daysSinceLastSession_F)

NR <- rbind(nrt,nrf) %>%
  arrange(dt) %>%
  rename(date = dt)


library(bayesAB)


nr_new <- NR %>%
  filter(userType == "New Visitor")
nr_rtn  <-  NR %>%
  filter(userType == "Returning Visitor")

AB_new_rtn <- bayesTest(
  nr_new$sales,
  nr_rtn$sales,
  priors = c("shape" = 9, "rate" = 3),
  distribution = 'poisson'
```

```
)


plot(AB_new_rtn)


library(bayesAB)

nr_short <- NR %>%
  filter(daysSinceLastSession <= 2)
nr_long <- NR %>%
  filter(daysSinceLastSession > 2)



AB_delay <- bayesTest(
  nr_short$sales,
  nr_long$sales,
  priors = c("shape" = 9, "rate" = 3),

  distribution = 'poisson'
)

plot(AB_delay)


library(bayesAB)

nr_new <- NR %>% filter(userType ==
                            "New Visitor" &
                            daysSinceLastSession <= 2)
nr_rtn <-   NR %>%
  filter(daysSinceLastSession > 2)
nr_short <- NR %>%
  filter(daysSinceLastSession <= 2)
nr_long <- NR %>%
```

```
  filter(daysSinceLastSession > 2)

AB_1st <- bayesTest(
  nr_new$sales,
  nr_rtn$sales,
  priors = c("shape" = 9, "rate" = 3),
  ## gives a distribution for number
  #of days to return centered on 3
  distribution = 'poisson'
)


AB_2nd <- bayesTest(
  nr_short$sales,
  nr_long$sales,
  priors = c("shape" = 9, "rate" = 3),

  distribution = 'poisson'
)


AB_combine <-
  combine(
    AB_1st,
    AB_2nd,
    f = `+`,
    params =
      c('Lambda', 'Lambda'),
    newName =
      'Combined Sales Posterior'
  )


plot(AB_combine)
```

12.8. R Code for Chapter 8

```r
# Load necessary libraries
library(ggplot2)
library(gridExtra)
library(knitr)
library(kableExtra)

# Define a function to calculate
#log-normal parameters,
#plot PDF and CDF, and compute NPV
plot_lognormal_npv <-
  function(
    mean_life_expectancy,
    sd_life_expectancy,
    color,
    discount_rate = 0.05,
    prior_mean, prior_sd) {
  # Convert mean and sd to log-normal parameters
  mu <-
    log(mean_life_expectancy^2 /
          sqrt(sd_life_expectancy^2 +
                 mean_life_expectancy^2))
  sigma <-
    sqrt(log(1 + (sd_life_expectancy^2 /
                    mean_life_expectancy^2)))

  # Generate a sequence of ages
  ages <- seq(0, 120, by = 0.1)

  # Calculate the probability density
  #function (PDF) and cumulative density function (CDF)
  pdf_values <- dlnorm(ages,
                       meanlog = mu,
                       sdlog = sigma)
  cdf_values <- plnorm(ages,
                       meanlog = mu,
                       sdlog = sigma)
```

```r
# Calculate NPV of £1,000,000 payout
#at the time of death
npv_values <- 10000 *
  exp(-discount_rate * ages) * pdf_values
npv <- sum(npv_values) *
  (ages[2] - ages[1])
# Integrate over the age range

# Bayesian analysis manually
data <- rlnorm(length(ages),
               meanlog = mu,
               sdlog = sigma)
n <- length(data)
data_mean <- mean(data)
data_var <- var(data)

# Calculate posterior mean and variance
post_var <- 1 /
  (1 / prior_sd^2 + n / data_var)
post_mean <- post_var *
  (prior_mean / prior_sd^2 + n *
     data_mean / data_var)
posterior_sd <- sqrt(post_var)

# Calculate NPV from posterior values
posterior_npv_values <-
  10000 * exp(-discount_rate * ages) *
  dlnorm(ages,
         meanlog = log(post_mean),
         sdlog = posterior_sd)
posterior_npv <-
  sum(posterior_npv_values) *
  (ages[2] - ages[1])

# Create a data frame for plotting
plot_data <- data.frame(
  Age = ages,
  PDF = pdf_values,
```

```r
    CDF = cdf_values)

# Plot the PDF
pdf_plot <- ggplot(plot_data,
                   aes(x = Age, y = PDF)) +
  geom_line(color = color) +
  ggtitle(paste("Age at Death")) +
  xlim(0,25) +
  xlab("Age") +
  ylab("Probability Density") +
  theme_minimal()

# Plot the CDF
cdf_plot <- ggplot(plot_data, aes(
  x = Age,
  y = CDF)) +
  geom_line(color = color) +
  ggtitle(paste("Age at Death")) +
  xlab("Age") +
  ylab("Cumulative Probability") +
      xlim(0,25) +
  theme_minimal()

  list(pdf_plot = pdf_plot,
       cdf_plot = cdf_plot,
       npv = npv,
       posterior_npv = posterior_npv)
}


# Scenario 2: Average age of
#death is 15 with standard deviation 10
scenario2 <-
  plot_lognormal_npv(mean_life_expectancy = 15,
                     sd_life_expectancy = 10,
                     color = "black",
                     prior_mean = 20,
                     prior_sd = 10)
```

```r
# Create a data frame for the results
results <- data.frame(
  Scenario = c("Dried Food", "Fresh Food"),
#  Mean_Age = c(5, 15),
#  SD_Age = c(10, 10),
  NPV =
  round(c(scenario1$npv,
          scenario2$npv), 2),
  Posterior_NPV =
  round(c(scenario1$posterior_npv,
          scenario2$posterior_npv), 2)
)


# Display the results in a table
kable(
  results,
  caption =
    "Comparative Actuarial
Costs of Pet Insurance",
  col.names = c(
    "Diet",
#    "Mean Age",
#    "SD Age",
    "Frequentist NP Cost of
    Insurance Claim ($)",
    "Bayesian Posterior NP
    Cost of Insurance Claim ($)"
  )
) %>%
  kable_styling(full_width = F,
                position = "center"
                ) %>%
  column_spec(1, width = "3cm") %>%
#  column_spec(2, width = "2cm") %>%
#  column_spec(3, width = "2cm") %>%
  column_spec(2, width = "3cm") %>%
  column_spec(3, width = "3cm")
```

```
# Arrange the plots


library(grid)
library(gridExtra)



grid.arrange(
  scenario1$pdf_plot,
  scenario2$pdf_plot,
  ncol = 2,
  top = textGrob(
    "Mortality Distributions for Dried Food
    (left) and Fresh Food (right)",
    gp = gpar(fontsize = 14, font = 3)
  )
)




grid.arrange(
  scenario1$cdf_plot,
  scenario2$cdf_plot,
  ncol = 2,
  top = textGrob(
    "Mortality Distributions for Dried Food
    (left) and Fresh Food (right)",
    gp = gpar(fontsize = 14, font = 3)
  )
)

library(tidyverse)
library(readr)
library(bayesAB)

set.seed(123)
```

```
girls_names <- read_csv(("girls_names.csv"),
                        col_names = FALSE)[, 2]


sdlog <- 2
meanlog <-

prof_A <- rpois(1000, 6) # mean 5.5
prof_B <- rpois(1000, 5) # mean 3.7
ron_rate <- rpois(1000, 7) # mean 6.0

prof_A <-ifelse(prof_A > 11, rpois(1000, 6), prof_A)
prof_B <-ifelse(prof_B > 1, rpois(1000, 5), prof_B)
ron_rate <-ifelse(ron_rate > 11, rpois(1000, 7),
                                      ron_rate)
prof_A <-ifelse(prof_A > 11, 10, prof_A)
prof_B <-ifelse(prof_B > 11,10, prof_B)
ron_rate <-ifelse(ron_rate > 11, 10, ron_rate)

## this tibble is the "ground truth"
#dating pool on MatchRate; both sides
#will in exceptional circumstances
#assign a rating higher than 10

dating_pool <- tibble(girls_names,
                      prof_A, prof_B,
                      ron_rate)

colnames(dating_pool) <-
  c("lady", "Profile_A",
    "Profile_B",
    "Ladies_rating")
write.csv(dating_pool,
          file="DATABASE_dating_pool.csv")
```

```r
# Melt the data for easier plotting with ggplot2
library(reshape2)
dating_pool <- melt(dating_pool)

# Plotting
ggplot(dating_pool, aes(x = value)) +
  geom_histogram(binwidth = 1, fill = "white",
                 color = "black") +
  # Adjust binwidth as needed
  facet_wrap(~ variable, scales = "free") +
  # Separate plot for each variable
  theme_minimal() +
  labs(x = "Ratings (higher is better)", y = "Count") +
  theme(legend.title = element_blank())
```

```r
dating_pool <-
  read.csv("DATABASE_dating_pool.csv")

# Bayesian A/B test
  bayes_test <- bayesAB::bayesTest(
    prof_A,
    prof_B,
    priors = c("shape" = 1, "rate" = 1),
    distribution = 'poisson'
  )

dating_pool <- data.frame(
  dating_pool$lady,
  bayes_test$posteriors$Lambda$A,
  bayes_test$posteriors$Lambda$B,
  dating_pool$Ladies_rating
)
colnames(dating_pool) <-
  c("lady",
    "Profile_A",
    "Profile_B",
    "Ladies rating")
```

```r
library(reshape2)
dating_pool <- melt(dating_pool)

# Plotting
ggplot(dating_pool, aes(x = value)) +
  geom_histogram(binwidth = .05,
                 fill = "white", color = "black") +
  # Adjust binwidth as needed
  facet_wrap(~ variable, scales = "free") +
  # Separate plot for each variable
  theme_minimal() +
  labs(x =
        "Ratings (higher is better)", y = "Count") +
  theme(legend.title = element_blank())


dating_pool <-
  read.csv("DATABASE_dating_pool.csv")


# Required libraries
library(tidyverse)

set.seed(123)

ron_benchmark <- 8
# the minimum rating for which
# either Ron or the ladies will consider dating
lady_benchmark <- 5


max_attempts <- 1000
nights <- tibble()

for (i in 1:1000) {
  i <- j <- 1

  # Profile A
  repeat {
```

```r
  tonight_pool <-
    dating_pool[sample(nrow(dating_pool),
                       30,
                       replace = TRUE), ]
  match_A <- subset(tonight_pool,
                    Profile_A > ron_benchmark &
                    Ladies_rating > lady_benchmark)
  if (nrow(match_A) > 0) {
    break
  }
  i <- i + 1
  if (i >= max_attempts) {
    print("Max attempts reached
          without finding a
          match for Profile A.")
    break
  }
}

# Profile B
repeat {
  tonight_pool <-
    dating_pool[sample(nrow(dating_pool),
                       30,
                       replace = TRUE), ]
  match_B <- subset(tonight_pool,
                    Profile_B > ron_benchmark &
                    Ladies_rating > lady_benchmark)
  if (nrow(match_B) > 0) {
    break
  }
  j <- j + 1
  if (j >= max_attempts) {
    print("Max attempts reached without
          finding a match for Profile B.")
    break
  }
}
```

```r
  nights <- rbind(nights, tibble(Prof_A = i, Prof_B = j))
}

colnames(nights) <- c("Profile A", "Profile B")

# Reshaping and plotting
nights_long <- pivot_longer(nights, everything(),
                            names_to = "profile",
                            values_to = "nights")
ggplot(nights_long, aes(x = nights)) +
  geom_histogram(binwidth = 1,
                 fill = "white", color = "black") +
  facet_wrap(~profile, scales = "free") +
  theme_minimal() +
  labs(x = "Number of Nights for
       Ron to Meet his Match",
       y = "How Often (Number of
       Nights out of 1000)") +
  theme(legend.title = element_blank()) +
  scale_y_continuous(trans = 'log10')
```

12.9.   R Code for Chapter 9

```r
library(wordcloud)
library(tidyverse)
library(tidytext)
library(tokenizers)
library(tm)
library(readr)
library(knitr)

reviews <- read_csv("DATABASE_reviews.csv") %>%
  select(UserId, Score, Text)

set.seed(123)
mkt <- sample(seq_len(nrow(reviews)),
```

```
                  size = nrow(reviews) / 2)
A <- reviews %>% filter(Score < 4)
B <- reviews %>% filter(Score >= 4)

# write.csv(A, B, reviews, 'NLP_reviews.csv')
```

NLP Analysis

```
# Load necessary libraries
library(dplyr)
library(tidyr)
library(tokenizers)
library(tidytext)

# Check and preprocess text data
preprocess_text <- function(data) {
  if (is.data.frame(data) && "Text" %in% names(data)) {
    data$Text <- as.character(data$Text)
    Encoding(data$Text) <- "UTF-8"
    text_stuff <- tokenize_words(data$Text) %>%
      unlist() %>%
      as.data.frame(stringsAsFactors = FALSE)
    colnames(text_stuff) <- "word"
    return(text_stuff)
  } else {
    stop("Input data must be a
        data frame with a 'Text' column")
  }
}

# Compute net sentiment
compute_net_sentiment <- function(text_stuff) {
  sentiments <- get_sentiments("nrc")
  net_sentiment <- text_stuff %>%
    inner_join(sentiments, by = "word") %>%
    count(sentiment) %>%
    pivot_wider(names_from = sentiment,
                values_from = n,
```

```
                 values_fill = list(n = 0)) %>%
     mutate(net_positive = positive - negative,
            proportion_positive =
                positive / negative - 1)
   return(net_sentiment)
}




# Process data A and B if they are correctly formatted
text_stuff_A <- preprocess_text(A)
net_sentiment_A <- compute_net_sentiment(text_stuff_A)

text_stuff_B <- preprocess_text(B)
net_sentiment_B <- compute_net_sentiment(text_stuff_B)

# Combine results and create a summary table
net_sentiment <- rbind(net_sentiment_A, net_sentiment_B)
product <- data.frame(Product = c("A", "B"),
                      stringsAsFactors = FALSE)
net_sentiment <- cbind(product, net_sentiment)

# Print the results
print(net_sentiment)


# Load necessary libraries
library(dplyr)
library(tidyr)
library(tokenizers)
library(tidytext)

# Check and preprocess text data
preprocess_text <- function(data) {
   if (is.data.frame(data) && "Text" %in% names(data)) {
     Encoding(data$Text) <- "UTF-8"
     text_stuff <- tokenize_words(data$Text) %>%
       unlist() %>%
       as.data.frame()
```

```
    colnames(text_stuff) <- "word"
    return(text_stuff)
  } else {
    stop("Input data must be a data
        frame with a 'Text' column")
  }
}


# Compute net sentiment
compute_net_sentiment <- function(text_stuff) {
  net_sentiment <- text_stuff %>%
    inner_join(get_sentiments("nrc"), by = "word") %>%
    count(sentiment) %>%
    spread(sentiment, n, fill = 0) %>%
    mutate(net_positive = positive - negative,
           proportion_positive =
               positive / negative - 1)
  return(net_sentiment)
}


# Process data A and B if they are correctly formatted
text_stuff_A <- preprocess_text(A)
net_sentiment_A <-
  compute_net_sentiment(text_stuff_A)


text_stuff_B <- preprocess_text(B)
net_sentiment_B <-
  compute_net_sentiment(text_stuff_B)


# Combine results and create a summary table
net_sentiment <-
  rbind(net_sentiment_A, net_sentiment_B)
product <- data.frame(Product = c("A", "B"))
net_sentiment <-
  cbind(product, net_sentiment)


# Print the results
print(net_sentiment)
```

```r
# Load necessary libraries
library(dplyr)
library(tidyr)
library(tokenizers)
library(tidytext)

# Function to check and preprocess text data
preprocess_text <- function(data) {
  # Validate that the input is a data frame
  if (!is.data.frame(data)) {
    stop("Input data must be a data frame.")
  }

  # Validate that the data frame contains a 'Text' column
  if (!"Text" %in% names(data)) {
    stop("Data frame must contain a 'Text' column.")
  }

  # Set the text encoding to UTF-8 and preprocess
  Encoding(data$Text) <- "UTF-8"
  text_stuff <- tokenize_words(data$Text) %>%
    unlist() %>%
    as.data.frame()
  colnames(text_stuff) <- "word"
  return(text_stuff)
}

# Function to compute net sentiment
compute_net_sentiment <- function(text_stuff) {
  net_sentiment <- text_stuff %>%
    inner_join(get_sentiments("nrc"), by = "word") %>%
    count(sentiment) %>%
    spread(sentiment, n, fill = 0) %>%
    mutate(net_positive = positive - negative,
           proportion_positive = positive / negative - 1)
  return(net_sentiment)
}
```

```r
# Process text data from A, ensuring it's properly
    formatted
if (is.data.frame(A) && "Text" %in% names(A)) {
  text_stuff_A <- preprocess_text(A)
  net_sentiment_A <-
    compute_net_sentiment(text_stuff_A)

  # Combine results and create a summary table
  net_sentiment <- rbind(net_sentiment_A)
  # Add more datasets as needed
  product <- data.frame(Product = c("A"))
  # Extend with more products as needed
  net_sentiment <- cbind(product, net_sentiment)

  # Print the results
  print(net_sentiment)
} else {
  print("Error: The data 'A' is not a
        data frame or lacks a 'Text' column.")
}
```

WordClouds

```r
A <- A[ceiling(runif(10000,0,nrow(A))),]
B <- B[ceiling(runif(10000,0,nrow(B))),]

cat("Wordcloud for Product A")

Encoding(A$Text) = "UTF-8"
text_stuff <- tokenize_words(A$Text) %>%
  unlist() %>%
  as.data.frame()

colnames(text_stuff) <- "word"

stuff_sentiment <-
  text_stuff %>%
  inner_join(get_sentiments("nrc"), by = "word")
```

```
stp <- get_stopwords()
stp <- rbind(stp, "good", "food", "love",
             "sweet", "chocolate", "bad", "sugar",
             "found", "weight", "smell", "money")

frequency_A <- text_stuff %>% anti_join(stp) %>%
  inner_join(stuff_sentiment) %>% count(word)
frequency_A  %>%
  with(wordcloud(
    word,
 colors = "black",
    rot.per = 0.15,
    n,
    max.words = 70
  ))

cat("Wordcloud for Product B")

Encoding(A$Text) = "UTF-8"
text_stuff <- tokenize_words(B$Text) %>%
  unlist() %>%
  as.data.frame()

colnames(text_stuff) <- "word"

stuff_sentiment <-
  text_stuff %>%
  inner_join(get_sentiments("nrc"),
             by = "word")

frequency_B <- text_stuff %>% anti_join(stp) %>%
  inner_join(stuff_sentiment) %>% count(word)
frequency_B  %>%
  with(wordcloud(
    word,
    colors = "black",
    rot.per = 0.15,
    n,
```

```
   max.words = 70
))
```

Top 10 Words for Each Sentiment

```
cat("Top 10 Words by Sentiment for Product A")

Encoding(A$Text) = "UTF-8"
text_stuff <- tokenize_words(A$Text) %>%
  unlist() %>%
  as.data.frame()

colnames(text_stuff) <- "word"


stuff_sentiment <-
text_stuff %>% anti_join(stp) %>%
  inner_join(get_sentiments("nrc"), by = "word")

text_counts <- stuff_sentiment %>%
count(word, sentiment, sort = TRUE) %>%
ungroup()

text_counts %>%
  group_by(sentiment) %>%
  top_n(10) %>%
  ungroup() %>%
  mutate(word = reorder(word, n)) %>%
  ggplot(aes(word, n, fill = sentiment)) +
  scale_fill_grey(start = 0, end = .2) +
  theme_minimal() +
  # Optional: starts with a minimal theme
  theme(axis.text.
        x = element_text(angle = 45, hjust = 1)) +
  geom_col(show.legend = FALSE) +
  facet_wrap(~ sentiment, scales = "free_y") +
  labs(y = "Contribution to sentiment", x = NULL) +
  coord_flip()
```

```
cat("Top 10 Words by Sentiment for Product B")

Encoding(B$Text) = "UTF-8"
text_stuff <- tokenize_words(B$Text) %>%
  unlist() %>%
  as.data.frame()

colnames(text_stuff) <- "word"


stuff_sentiment <-
text_stuff %>% anti_join(stp) %>%
  inner_join(get_sentiments("nrc"), by = "word")

text_counts <- stuff_sentiment %>%
count(word, sentiment, sort = TRUE) %>%
ungroup()

text_counts %>%
  group_by(sentiment) %>%
  top_n(10) %>%
  ungroup() %>%
  mutate(word = reorder(word, n)) %>%
  ggplot(aes(word, n, fill = sentiment)) +
  scale_fill_grey(start = 0, end = .2) +
  theme_minimal() +
  # Optional: starts with a minimal theme
  theme(axis.text.x = element_text(
    angle = 45, hjust = 1)
    ) +
  geom_col(show.legend = FALSE) +
  facet_wrap(~ sentiment, scales = "free_y") +
  labs(y = "Contribution to sentiment", x = NULL) +
  coord_flip()


library(tidyverse)
library(tidytext)
```

```
library(tokenizers)
library(tm)
library(readr)
library(knitr)


Encoding(A$Text) = "UTF-8"
text_stuff <- tokenize_words(A$Text) %>%
  unlist() %>%
  as.data.frame()
colnames(text_stuff) <- "word"
stuff_sentiment_A <-
  text_stuff %>%
  inner_join(get_sentiments("nrc"), by = "word") %>%
  count(sentiment, word)

sum_sent <- stuff_sentiment_A %>%
  group_by(sentiment) %>%
  summarize(mu = mean(n), sig = sd(n))
sent_stats_A <-
  data.frame(sum_sent$sentiment,
             sum_sent$mu ,
             sum_sent$sig)
sent_stats_A


Encoding(B$Text) = "UTF-8"
text_stuff <- tokenize_words(B$Text) %>%
  unlist() %>%
  as.data.frame()
colnames(text_stuff) <- "word"
stuff_sentiment_B <-
  text_stuff %>%
  inner_join(get_sentiments("nrc"),
             by = "word") %>%
  count(sentiment, word)

sum_sent <- stuff_sentiment_B %>%
```

```r
  group_by(sentiment) %>%
  summarize(mu = mean(n),
            sig = sd(n))
sent_stats_B <-
  data.frame(sum_sent$sentiment,
             sum_sent$mu ,
             sum_sent$sig)
sent_stats_B


library(bayesAB)

snt <- unique(stuff_sentiment_A$sentiment)
sent_stats <- data.frame()
for (i in snt) {
  sent_A <- stuff_sentiment_A %>%
    filter(sentiment == i)
  sent_B <- stuff_sentiment_B %>%
    filter(sentiment == i)
  AB1 <- bayesTest(
    sent_A$n,
    sent_B$n,
    priors = c(
      "mu" = 0,
      "lambda" = 1,
      "alpha" = 1,
      "beta" = 1
    ),
    distribution = 'normal'
  )
cat("Sentiment = ", i)
#print(AB1)
#summary(AB1)
#print(plot(AB1))

# Load the necessary library
library(ggplot2)
```

```r
# Example data vectors
Choice_A <- AB1$posteriors$Mu$A
# Generate some normal data
Choice_B <- AB1$posteriors$Mu$B
# Generate some normal data

# Create a data frame for ggplot
data <- data.frame(value = c(data1, data2),
                   group =
                       factor(rep(c("Choice A",
                                    "Choice B"),
                                  each = 100)))

# Create the density plot
p <- ggplot(data, aes(x = value, group = group)) +
  geom_density(aes(linetype = group), size = 1) +
  # Use different linetypes for each group
  scale_linetype_manual(values = c("solid", "dashed")) +
  # Assign solid and dashed lines
  facet_wrap(~ group, scales = "free_y") +
  # Facet by group, allowing free scales on the y-axis
  labs(title = "Density Plot by Group",
       x = i, y = "Density") +
  theme_minimal() +
  # Using a minimal theme for better visibility
  theme(text = element_text(color = "black"),
        # Ensure text is black for contrast
        panel.grid.major = element_blank(),
        # Remove major grid lines
        panel.grid.minor = element_blank(),
        # Remove minor grid lines
        panel.background = element_blank(),
        # Remove panel background
        plot.background = element_blank(),
        # Transparent plot background
        strip.background = element_blank(),
        # Transparent facet strip background
        strip.text = element_text(color = "black"))
```

```
print(p)


}


# Load the necessary library
library(ggplot2)

# Example data vectors
data1 <- rnorm(100, mean = 50, sd = 10)
# Generate some normal data
data2 <- rnorm(100, mean = 60, sd = 15)
# Generate some normal data with different parameters

# Create a data frame for ggplot
data <- data.frame(value = c(data1, data2), group =
                   factor(rep(c("Data1",
                               "Data2"),
                           each = 100)))

# Create the density plot
ggplot(data) +
  geom_density(aes(x = value, fill = group,
                 color = group), alpha = 0.5) +
  scale_fill_manual(values = c("blue", "red")) +
  # Set custom fill colors for the groups
  scale_color_manual(values = c("blue", "red")) +
  # Set custom line colors for the groups
  labs(title = "Density Plot",
      x = "Value",
      y = "Density") +
  theme_minimal()
# Using a minimal theme for better visibility


# Load the necessary library
library(ggplot2)
```

```r
# Example data vectors
Choice_A <- AB1$posteriors$Mu$A
# Generate some normal data
Choice_B <- AB1$posteriors$Mu$B
# Generate some normal data with different parameters

# Create a data frame for ggplot
data <- data.frame(value = c(data1, data2),
                   group =
                       factor(rep(c("Choice A", "Choice B"),
                                  each = 100)))

# Create the density plot
ggplot(data, aes(x = value, group = group)) +
  geom_density(aes(linetype = group), size = .3) +
  # Different linetypes for each group
  scale_linetype_manual(values =
                            c("solid", "dashed")) +
  # Assign solid and dashed lines
  labs(title = "Density Plot",
       x = "Value",
       y = "Density") +
  theme_minimal() +
  # Using a minimal theme for better visibility
  theme(text = element_text(color = "black"),
        # Ensure text is black for contrast
        panel.grid.major = element_blank(),
        # Remove major grid lines
        panel.grid.minor = element_blank(),
        # Remove minor grid lines
        panel.background = element_blank(),
        # Remove panel background
        plot.background = element_blank())
# Transparent plot background


library(bayesAB)
library(readr)
```

```
priors <-
  read_csv(
    "DATABASE_priors_for_NLP_case_study.csv"
    )[1:10,c(2,5:8)]



snt <- unique(stuff_sentiment_A$sentiment)
sent_stats <- data.frame()
for (i in c(
  "anger",
  joy,
  "anticipation",
  "disgust",
  "fear",
  "sadness",
  "surprise",
  "trust",
  "negative",
  "positive"
  )) {
  sent_A <-
    stuff_sentiment_A %>% filter(sentiment == i)
  sent_B <-
    stuff_sentiment_B %>% filter(sentiment == i)
  prr <- priors %>% filter(Sentiment == i)
  AB1 <- bayesTest(
    sent_A$n,
    sent_B$n,
    priors = c(
      "mu" = prr$mu,
      "lambda" = prr$lambda,
      "alpha" = prr$alpha,
      "beta" = prr$beta
    ),
    distribution = 'normal'
  )
```

```
cat("Sentiment = ", i)
#print(AB1)
#summary(AB1)
#print(plot(AB1))

# Load the necessary library
library(ggplot2)

# Example data vectors
data1 <- AB1$posteriors$Mu$A
# Generate some normal data
data2 <- AB1$posteriors$Mu$A
# Generate some normal data with different parameters

# Create a data frame for ggplot
data <- data.frame(value = c(data1, data2),
                    group =
                        factor(rep(c("Data1",
                                     "Data2"),
                                   each = 100)))

# Create the density plot
ggplot(data, aes(x = value,
                 fill = group,
                 alpha = 0.5)) +
  geom_density(adjust = 1.5) +
  # Adjust parameter on how smooth you want the
    curve to be
  scale_fill_manual(values = c("blue", "red")) +
  # Set custom colors for the groups
  scale_alpha_manual(values = c(0.5, 0.5)) +
  # Set transparency level
  labs(title = "Density Plot",
       x = "Value",
       y = "Density") +
  theme_minimal()
# Using a minimal theme for better visibility
```

```
}


library(kableExtra)
library(tidyverse)
library(tidytext)
library(tokenizers)
library(tm)
library(readr)
library(knitr)


reviews <- read_csv("DATABASE_reviews.csv") %>%
  select(UserId, Score, Text)

set.seed(123)
mkt <- sample(seq_len(nrow(reviews)),
              size = nrow(reviews) / 2)
A <- reviews %>% filter(Score < 4)
B <- reviews %>% filter(Score >= 4)

Encoding(A$Text) = "UTF-8"
text_stuff <- tokenize_words(A$Text) %>%
  unlist() %>%
  as.data.frame()
colnames(text_stuff) <- "word"
stuff_sentiment_A <-
  text_stuff %>%
  inner_join(get_sentiments("nrc"), by = "word") %>%
  count(sentiment, word)

sum_sent <-
  stuff_sentiment_A %>%
  group_by(sentiment) %>%
  summarize(mu = mean(n), sig = sd(n))
sent_stats_A <-
  data.frame(sum_sent$sentiment,
             sum_sent$mu ,
```

```
                    sum_sent$sig)
sent_stats_A %>% kbl(
  caption="Sentiment A Statistics")


Encoding(B$Text) = "UTF-8"
text_stuff <- tokenize_words(B$Text) %>%
  unlist() %>%
  as.data.frame()
colnames(text_stuff) <- "word"
stuff_sentiment_B <-
  text_stuff %>%
  inner_join(get_sentiments("nrc"), by = "word")  %>%
  count(sentiment, word)

sum_sent <- stuff_sentiment_B %>%
  group_by(sentiment) %>%
  summarize(mu = mean(n), sig = sd(n))
sent_stats_B <-
  data.frame(sum_sent$sentiment,
             sum_sent$mu ,
             sum_sent$sig)
sent_stats_B %>% kbl(caption="Sentiment B Statistics")
```

12.10. R Code for Chapter 10

```
library(bayesAB)
library(tidyverse)

bs <- read_csv("kildare_bs.csv")


pre_theft <-  bs %>%
  filter(Year < 2015) %>%
  select(Equity) %>%
  unlist() %>%
```

```
  as.integer()
post_theft <- bs %>%
  filter(Year >= 2015) %>%
  select(Equity) %>%
  unlist() %>%
  as.integer()

AB_NPV <- bayesTest(
    pre_theft,
    post_theft,
    priors = c("shape" = 10,
               "rate" = 10/mean(bs$Equity)),
    distribution = 'poisson'
  )

plot(AB_NPV)
```

12.11. R Code for Chapter 11

```
library(readr)
library(bayesAB)
library(tidyverse)



fin <- read_csv("kildare_inc_stmt.csv")
deflator <- .04  # assume 4% annual inflation

efin <- dfin <- fin %>%
  mutate(
    deflate = (1 + deflator)^(Year-2023),
    Revenue = trunc(Revenue / deflate),
    CGS = trunc(CGS / deflate),
    GP = trunc(GP / deflate),
    RD = trunc(RD / deflate),
    OE = trunc(OE / deflate),
```

```
    NI = trunc(NI / deflate)
  )


pre_theft <-  efin %>%
  filter(Year < 2015) %>%
  select(NI) %>%
  unlist() %>%
  as.numeric()
post_theft <- efin %>%
  filter(Year >= 2015) %>%
  select(NI) %>%
  unlist() %>%
  as.numeric()

AB_profit <- bayesTest(
    pre_theft,
    post_theft,
    priors = c("shape" = 10, "rate" = 10/mean(efin$NI)),
    distribution = 'poisson'
  )

plot(AB_profit)
```

Bandit with Bernoulli distributed data and Thompson sampling

```
library(ggplot2)
library(gridExtra)

# Function to simulate bandit arms
# with Bernoulli rewards

make_bandits <- function(params) {
  pull <- function(arm) {
    function() {
      rbinom(1, 1, params[arm + 1])
    }
  }
}
```

```
  list(pull = pull,
       num_bandits = length(params))
}


# Bayesian strategy for bandit selection
bayesian_strategy <-
  function(pull, num_bandits) {
  num_rewards <- numeric(num_bandits)
  num_trials <- numeric(num_bandits)

  list(
    play = function() {
      choice <-
        which.max(rbeta(
          num_bandits,
          2 + num_rewards,
          2 + num_trials - num_rewards)) - 1
      reward <- pull(choice)()
      num_rewards[choice + 1] <-
        num_rewards[choice + 1] + reward
      num_trials[choice + 1] <-
        num_trials[choice + 1] + 1
      list(choice = choice,
           reward = reward,
           num_rewards = num_rewards,
           num_trials = num_trials)
    }
  )
}


# Plot posterior distributions
plot_posteriors <- function(num_rewards,
                            num_trials,
                            n,
                            title_suffix) {
  x <- seq(0, 1, length.out = 100)
  data <- expand.grid(x = x,
                      Bandit = seq_along(num_rewards))
```

```r
  densities <- mapply(
    function(a, b) dbeta(x, a + 2, b - a + 2),
    a = num_rewards,
    b = num_trials,
    SIMPLIFY = FALSE
  )


  # Flatten the list
  # and attach it to the dataframe
  data$y <- unlist(densities)

  ggplot(data, aes(x = x,
                   y = y,
                   group = Bandit)) +
    geom_line(aes(linetype = as.factor(Bandit)),
              color = "grey20") +
    geom_area(aes(fill = as.factor(Bandit)),
              alpha = 0.3) +
    scale_linetype_manual(
      values = c("solid", "dashed", "dotted")) +
    scale_fill_manual(
      values = c("grey10", "grey30", "grey50")) +
    labs(title = paste('Posterior after',
                       n,
                       'pulls',
                       title_suffix),
         x = 'Probability',
         y = 'Density') +
    theme_minimal() +
    theme(legend.title = element_blank())
}

# Setup and play the game
bandit_setup <-
  make_bandits(c(0.2, 0.5, 0.7))
num_bandits <-
  bandit_setup$num_bandits
play_function <-
```

```
  bayesian_strategy(bandit_setup$pull,
                    num_bandits)$play

pulls_to_plot <- c(1, 2, 10, 20, 50, 100)

# Plot at specific points
plots <- lapply(pulls_to_plot, function(n) {
  for (i in 1:n) {
    result <- play_function()
  }
  plot_posteriors(result$num_rewards,
                  result$num_trials,
                  n,
                  ifelse(n > 1, 'pulls', 'pull'))
})

# Display plots
gridExtra::grid.arrange(grobs = plots, ncol = 2)
```

Bandit code with Count Distributions

```
library(ggplot2)
library(reshape2)  # For melting data frames
library(gridExtra) # For arranging plots

set.seed(123)

# Create bandits with Poisson-distributed rewards
make_bandits <- function(params) {
  pull <- function(arm) {
    function() {
      rpois(1, lambda = params[arm + 1])
    }
  }

  list(pull = pull, num_bandits = length(params))
}
```

```r
# Bayesian strategy, Gamma priors
bayesian_strategy <- function(pull, num_bandits) {
  num_rewards <- rep(1, num_bandits)
  num_trials <- rep(1, num_bandits)

  list(
    play = function() {
      choice <-
        which.max(rgamma(num_bandits,
                              shape = num_rewards,
                              rate = num_trials)) - 1
      reward <- pull(choice)()
      num_rewards[choice + 1] <-
        num_rewards[choice + 1] + reward
      num_trials[choice + 1] <-
        num_trials[choice + 1] + 1
      list(
        choice = choice,
        reward = reward,
        num_rewards = num_rewards,
        num_trials = num_trials
      )
    }
  )
}


# Function to plot the posterior
plot_posteriors <-
  function(num_rewards,
           num_trials,
           n) {
  x <- seq(0, 10, length.out = 100)
  plot_data <- data.frame(x = x)

  # Compute densities
  for (i in seq_along(num_rewards)) {
    plot_data[[paste0("Bandit", i)]] <-
      dgamma(x, shape = num_rewards[i],
```

```
                    rate = num_trials[i])
}

# Reshape for plotting
plot_data_melted <- melt(
  plot_data,
  id.vars = "x",
  variable.name = "Bandit",
  value.name = "Density"
)

# Plot configurations
linetypes <- c("solid", "dashed", "dotted")
colors <- c("black", "black", "black")

p <- ggplot(plot_data_melted, aes(x = x,
                                  y = Density,
                                  group = Bandit)) +
  geom_line(aes(linetype = Bandit),
            size = .4,
            color = colors[as.integer(as.factor(
              plot_data_melted$Bandit))]) +
  geom_area(aes(fill = Bandit), alpha = 0.1) +
  scale_linetype_manual(values = linetypes) +
  scale_fill_manual(values = colors) +
  labs(
    title = sprintf("Posterior
                     distributions
                     after %d pulls",
                     n),
    x = "Reward size",
    y = "Density"
  ) +
  theme_minimal() +
  theme(legend.title = element_blank())

  return(p)
}
```

```r
# Initialize bandits and play function
bandit_setup <- make_bandits(c(4, 4.5, 5))
num_bandits <- bandit_setup$num_bandits
play_function <- bayesian_strategy(
  bandit_setup$pull,
  num_bandits)$play

# Collect plots for specific pulls
pulls_to_plot <- c(1, 2, 10, 20, 50, 100)
plots <- list()

for (i in 1:max(pulls_to_plot)) {
  result <- play_function()
  if (i %in% pulls_to_plot) {
    plots[[length(plots) + 1]] <-
      plot_posteriors(
        result$num_rewards,
        result$num_trials, i)
  }
}

# Arrange and display all plots in a grid
do.call(grid.arrange, c(plots, ncol = 2))
```

Shiny example for Real-time interactive implementation

```r
library(shiny)
library(ggplot2)
library(gridExtra)

# Define UI for application
ui <- fluidPage(
  titlePanel("Bayesian Bandit Simulation"),
  sidebarLayout(
    sidebarPanel(
      helpText("Simulate bandit arms with
                Bayesian updates on
                reward probabilities."),
```

```
    numericInput("bandit1",
                "Probability of
                reward for Bandit 1:",
                0.2,
                min = 0,
                max = 1),
    numericInput("bandit2",
                "Probability of
                reward for Bandit 2:",
                0.5,
                min = 0,
                max = 1),
    numericInput("bandit3",
                "Probability of
                reward for Bandit 3:",
                0.7,
                min = 0,
                max = 1),
    sliderInput("pulls", "Number of pulls:",
                min = 1,
                max = 500,
                value = 100),
    actionButton("simulate", "Simulate")
  ),
  mainPanel(
    plotOutput("plots")
  )
)
)

# Define server logic
server <- function(input, output) {
  observeEvent(input$simulate, {
    isolate({
      params <- c(input$bandit1,
                  input$bandit2,
                  input$bandit3)
      num_bandits <- length(params)
```

```
num_rewards <- numeric(num_bandits)
num_trials <- numeric(num_bandits)

pull <- function(arm) {
  function() {
    rbinom(1, 1, params[arm + 1])
  }
}

for (i in 1:input$pulls) {
  choice <- which.max(rbeta(
    num_bandits,
    2 + num_rewards,
    2 + num_trials - num_rewards)) - 1
  reward <- pull(choice)()
  num_rewards[choice + 1] <-
    num_rewards[choice + 1] + reward
  num_trials[choice + 1] <-
    num_trials[choice + 1] + 1
}

plots <-
  lapply(seq_len(num_bandits),
         function(bandit) {
  data <- data.frame(
    x = seq(0, 1, length.out = 100),
    y = dbeta(seq(0, 1,
                  length.out = 100),
              num_rewards[bandit] + 2,
              num_trials[bandit] -
                num_rewards[bandit] + 2)
  )
  ggplot(data, aes(x = x, y = y)) +
    geom_line() +
    geom_area(fill = "gray", alpha = 0.3) +
    labs(title = sprintf(
      "Posterior for Bandit %d after %d pulls",
      bandit, input$pulls),
```

```
              x = 'Probability', y = 'Density') +
          theme_minimal()
      })

      output$plots <- renderPlot({
        do.call(gridExtra::grid.arrange,
              c(plots, ncol = 1))
      })
    })
  })
}

# Run the application
shinyApp(ui = ui, server = server)
```

References

[1] Stolberg, M. (2006). Inventing the randomized double-blind trial: The nuremberg salt test of 1835. *Journal of the Royal Society of Medicine*, *99*(12), 642–643.

[2] Hopkins, C. C. (1968). *Scientific Advertising*. New Line Publishing.

[3] Box, J. F. (1987). Guinness, gosset, fisher, and small samples. *Statistical Science*, *2*(1), 45–52.

[4] Akerlof, G. A. & Shiller, R. J. (2015). *Phishing for Phools: The Economics of Manipulation and Deception*. Princeton University Press.

[5] Duhigg, C. (2012). *The Power of Habit: Why We Do What We Do in Life and Business* (Vol. 34). Random House.

[6] Neyman, J. & Pearson, E. S. (1933). IX. On the problem of the most efficient tests of statistical hypotheses. *Philosophical Transactions of the Royal Society of London. Series A, Containing Papers of a Mathematical or Physical Character*, *231*(694–706), 289–337.

[7] Neyman, J. (1970). A glance at some of my personal experiences in the process of research. *Scientists at Work*, 148–164.

[8] Aschwanden, C. (2015). Not even scientists can easily explain p-values. *FiveThirtyEight*. Retrieved from https://fivethirtyeight. com/features/not-even-scientists-can-easily-explain-p-values/.

[9] Hubbard, R. & Lindsay, R. M. (2008). Why p values are not a useful measure of evidence in statistical significance testing. *Theory & Psychology*, *18*(1), 69–88.

[10] Munafò, M. R., Nosek, B. A., Bishop, D. V., Button, K. S., Chambers, C. D., Percie du Sert, N., ... Ioannidis, J. (2017). A manifesto for reproducible science. *Nature Human Behaviour*, *1*(1), 1–9.

[11] Wasserstein, R. L. & Lazar, N. A. (2016). The ASA statement on p-values: Context, process, and purpose. *The American Statistician*. Taylor & Francis.

[12] Camerer, C. F., Dreber, A., Holzmeister, F., Ho, T.-H., Huber, J., Johannesson, M., *et al.* (2018). Evaluating the replicability of social science experiments in nature and science between 2010 and 2015. *Nature Human Behaviour, 2*(9), 637–644.

[13] Ioannidis, J. P. (2005). Why most published research findings are false. *PLoS medicine, 2*(8), e124.

[14] Gronau, Q. F., Raj, K., & Wagenmakers, E.-J. (2019). Informed bayesian inference for the *A/B* test. arXiv preprint arXiv:1905.02068.

[15] Meyer, B. D. (1995). Natural and quasi-experiments in economics. *Journal of Business & Economic Statistics, 13*(2), 151–161.

[16] Keysers, C., Gazzola, V., & Wagenmakers, E.-J. (2020). Using bayes factor hypothesis testing in neuroscience to establish evidence of absence. *Nature Neuroscience, 23*(7), 788–799.

[17] Robinson, G. K. (2018). What properties might statistical inferences reasonably be expected to have?—crisis and resolution in statistical inference. *The American Statistician.*

[18] Altman, D. G. & Bland, J. M. (1995). Statistics notes: Absence of evidence is not evidence of absence. *Bmj, 311*(7003), 485.

[19] Berger, J. O. & Wolpert, R. L. (1988). The likelihood principle. In *IMS.*

[20] Wagenmakers, E.-J. (2007). A practical solution to the pervasive problems of p values. *Psychonomic Bulletin & Review, 14*(5), 779–804.

[21] Wagenmakers, E.-J., Love, J., Marsman, M., Jamil, T., Ly, A., Verhagen, J., *et al.* (2018). Bayesian inference for psychology. Part II: Example applications with JASP. *Psychonomic Bulletin & Review, 25*, 58–76.

[22] Wagenmakers, E.-J., Marsman, M., Jamil, T., Ly, A., Verhagen, J., Love, J., *et al.* (2018). Bayesian inference for psychology. Part i: Theoretical advantages and practical ramifications. *Psychonomic Bulletin & Review, 25*, 35–57.

[23] Van der Lee, J., Wesseling, J., Tanck, M., & Offringa, M. (2008). Efficient ways exist to obtain the optimal sample size in clinical trials in rare diseases. *Journal of Clinical Epidemiology, 61*(4), 324–330.

[24] Roe-Sepowitz, D. E., Hickle, K. E., Loubert, M. P., & Egan, T. (2011). Adult prostitution recidivism: Risk factors and impact of a diversion program. *Journal of Offender Rehabilitation, 50*(5), 272–285.

[25] Zondervan-Zwijnenburg, M., Peeters, M., Depaoli, S., & Van de Schoot, R. (2017). Where do priors come from? Applying guidelines to construct informative priors in small sample research. *Research in Human Development, 14*(4), 305–320.

[26] Rocchetti, L., Amato, A., Fonti, V., Ubaldini, S., De Michelis, I., Kopacek, B., ... Beolchini, F. (2015). Cross-current leaching of indium from end-of-life LCD panels. *Waste Management, 42,* 180–187.

[27] Lee, R. van der, Ellemers, N., Scheepers, D., & Rutjens, B. T. (2017). In or out? How the perceived morality (vs. Competence) of prospective group members affects acceptance and rejection. *European Journal of Social Psychology, 47*(6), 748–762.

[28] Boomsma, A. & Hoogland, J. J. (2001). The robustness of LISREL modeling revisited. *Structural equation models: Present and Future. A Festschrift in Honor of Karl Jöreskog, 2*(3), 139–168.

[29] Hox, J. J. & Maas, C. J. (2001). The accuracy of multilevel structural equation modeling with pseudobalanced groups and small samples. *Structural equation Modeling, 8*(2), 157–174.

[30] Jeffreys, H. (1946). An invariant form for the prior probability in estimation problems. *Proceedings of the Royal Society of London. Series A. Mathematical and Physical Sciences, 186*(1007), 453–461.

[31] Bernardo, J. M. (1979). Reference posterior distributions for bayesian inference. *Journal of the Royal Statistical Society Series B: Statistical Methodology, 41*(2), 113–128.

[32] Doeswijk, R., Lam, T., & Swinkels, L. (2020). Historical returns of the market portfolio. *The Review of Asset Pricing Studies, 10*(3), 521–567.

[33] Brandão, L. E., Dyer, J. S., & Hahn, W. J. (2005). Using binomial decision trees to solve real-option valuation problems. *Decision Analysis, 2*(2), 69–88.

[34] Hahn, W. J. & Dyer, J. S. (2008). Discrete time modeling of mean-reverting stochastic processes for real option valuation. *European Journal of Operational Research, 184*(2), 534–548.

[35] Tan, B., Anderson, E., Dyer, J., & Parker, G. (2009). Using binomial decision trees and real options theory to evaluate system dynamics models of risky projects. In *Proceedings of the 27th International Conference of the System Dynamics Society* (pp. 26–30).

[36] Westland, J. C. (2022). Information loss and bias in likert survey responses. *PloS One, 17*(7), e0271949.

[37] Dolley, J. C. (1934). The price-effect of stock right issues. *The Journal of Business of the University of Chicago, 7*(2), 133–160.

[38] Myers, J. H. & Bakay, A. J. (1948). Influence of stock split-ups on market price. *Harvard Business Review, 26*(2), 251–255.

[39] Barker, C. A. (1957). *Stock Splits in a Bull Market.* Harvard Business Review.

[40]   Barker, C. A. (1958). Evaluation of stock dividends. *Harvard Business Review, 36*(4), 99–114.

[41]   Barker, C. A. (1956). Effective stock splits. *Harvard Business Review, 34*(1), 101–106.

[42]   Ashley, J. W. (1962). Stock prices and changes in earnings and dividends: Some empirical results. *Journal of Political Economy, 70*(1), 82–85.

[43]   Ball, R. & Brown, P. (2013). An empirical evaluation of accounting income numbers. In *Financial accounting and equity markets* (pp. 27–46). Routledge.

[44]   Fama, E. F. (1998). Market efficiency, long-term returns, and behavioral finance. *Journal of Financial Economics, 49*(3), 283–306.

[45]   Fama, E. F. & French, K. R. (1996). Multifactor explanations of asset pricing anomalies. *The Journal of Finance, 51*(1), 55–84.

[46]   Schwert, G. W. (1981). Using financial data to measure effects of regulation. *The Journal of Law and Economics, 24*(1), 121–158.

[47]   Mitchell, M. L. & Netter, J. M. (1993). The role of financial economics in securities fraud cases: Applications at the securities and exchange commission. *Business Law, 49*, 545.

[48]   Zeuthen, F. (n.d.). JR hicks: Value and Capital. An Inquiry into Some Fundamental Principles of Economic Theory. Clarendon press, oxford 1939. XI+ 331 s. *Nationaløkonomisk Tidsskrift*.

[49]   Kothari, S. (2006). J B Warner "econometrics of event studies" in B Espen Eckbo (ed.) *Handbook of Corporate Finance: Empirical Corporate Finance Vol. A. North Holland, Elsevier*.

[50]   Brown, S. J. & Warner, J. B. (1980). Measuring security price performance. *Journal of Financial Economics, 8*(3), 205–258.

[51]   Brown, S. J. & Warner, J. B. (1985). Using daily stock returns: The case of event studies. *Journal of Financial Economics, 14*(1), 3–31.

[52]   MacKinlay, A. C. (1997). Event studies in economics and finance. *Journal of Economic Literature, 35*(1), 13–39.

[53]   Bubeck, S., Cesa-Bianchi, N., *et al.* (2012). Regret analysis of stochastic and nonstochastic multi-armed bandit problems. *Foundations and Trends® in Machine Learning, 5*(1), 1–122.

[54]   Scott, S. L. (2010). A modern bayesian look at the multi-armed bandit. *Applied Stochastic Models in Business and Industry, 26*(6), 639–658.

[55]   Davidson-Pilon, C. (2013). *Data Origami*. Retrieved from https:// dataorigami.net/2013/10/14/Multi-Armed-Bandits.html.

[56]   Agrawal, S. & Goyal, N. (2012). Analysis of thompson sampling for the multi-armed bandit problem. In *Conference on Learning Theory* (pp. 1–39). JMLR Workshop and Conference Proceedings.

Index